The Mathematical Theory of
ELASTICITY
&
The Mathematical Theory of
PLASTICITY

J. N. Goodier & P. G. Hodge, Jr.

Dover Publications, Inc.
Mineola, New York

Bibliographical Note

This Dover edition, first published in 2016, is an unabridged republication of the work originally published in 1958 by John Wiley & Sons, Inc., New York, as Volume I of the series "Surveys in Applied Mathematics." The books in this series were written as a joint project of the Office of Naval Research and Applied Mechanics Reviews.

Library of Congress Cataloging-in-Publication Data

Names: Goodier, J. N. (James Norman), 1905– | Goodier, J. N. (James Norman), 1905– Mathematical theory of elasticity. | Hodge, Philip Gibson, 1920– Mathematical theory of plasticity.
Title: Elasticity and plasticity : the mathematical theory of elasticity / J.N. Goodier. The mathematical theory of plasticity / P.G. Hodge, Jr.
Description: Dover edition. | Mineola, New York : Dover Publications, Inc., 2016. | Originally published: New York : Wiley, 1958. | Includes bibliographical references and index.
Identifiers: LCCN 2015038960| ISBN 9780486806044 | ISBN 0486806049
Subjects: LCSH: Elasticity. | Plasticity.
Classification: LCC QA931 .G6 2016 | DDC 531/.382–dc23 LC record available at http://lccn.loc.gov/2015038960

Manufactured in the United States by RR Donnelley
80604901   2016
www.doverpublications.com

PREFACE

These survey articles, as any of their kind, have the primary objective of accounting in a summary fashion for the state of the fields which they cover, as determined by contributing developments at various times and in many places. Considering the sustained growth of mathematics itself, as well as the intensive use of an ever-increasing number of its branches in diversified applied contexts, the availability of such surveys in selected areas is believed to fill a real need. The present articles are therefore aimed not so much at research specialists, actively contributing to the subjects discussed, as they are aimed at a broader, mathematically literate audience, looking for contemporary information on the important problems and results in these disciplines, whether it be for use in classroom and seminar, or for the sake of possible application to problems in other fields of science and engineering, or simply for reasons of personal interest.

The selection of the areas surveyed, as well as their coverage, was further guided by giving first consideration to developments of whose current state no comprehensive picture could be obtained by going only to the readily accessible literature in familiar languages. It is a unique distinction of the mathematical community to have never been taken in by the myth, now generally shattered, that Russian science merely followed the lead of the West—at a respectful distance. Mathematicians remained aware of the vigorous development during the postwar years of research in their field also in the Communist countries, and they knew of the steady stream of important results which it produced. An early start was made to overcome language and other communication difficulties. The American Mathematical Society was the first of this country's scientific societies to institute, with support from the Office of Naval Research, the systematic selection and translation of significant articles which appeared in inaccessible journals or unfamiliar languages. A further step was taken two years ago, when the Editorial Office of *Applied Mechanics Reviews*, again with Office of Naval Research support, initiated work on these surveys. It had become apparent by then that there existed major areas of modern mathematics and theoretical mechanics of whose current scope and fabric the Western literature alone—quite apart from missing individual recent results—conveyed only an incomplete and therefore inadequate picture. This is to be remedied in the areas covered by the present *Surveys in Applied Mathematics*.

Clearly, the success of this venture required that the surveys should be written by authorities in the respective fields, fully conversant with current research and abreast of the international literature. Special thanks must therefore be extended to the distinguished authors of these articles for having given their thought and time to the purpose at hand. The enlistment of their co-operation is due in large measure to the indefatigable leadership of the Editor of *Applied Mechanics Reviews* and his associates, as well as to the effective way in which the Midwest Research Institute and the Southwest Research Institute (after *Applied Mechanics Reviews* had moved there) jointly arranged for the conduct of the editorial work.

F. JOACHIM WEYL, Director
Mathematical Sciences Division
Office of Naval Research

Contents

THE MATHEMATICAL THEORY OF ELASTICITY

Article 1	Introduction. Scope and Intention	3
Article 2	Plane Stress and Plane Strain in the Isotropic Medium	4
Article 3	Holes and Fillets of Assignable Shapes. Approximate Conformal Mapping	8
Article 4	Reinforcement of Holes	11
Article 5	Mixed Boundary Value Problems. The Third Fundamental Problem in Two Dimensions	14
Article 6	Eigensolutions for Plane and Axisymmetric States	20
Article 7	Anisotropic Elasticity	21
Article 8	Thermal Stress. Elastic Waves Induced by Thermal Shock	26
Article 9	Three-Dimensional Contact Problems	29
Article 10	Wave Propagation. Traveling Loads and Sources of Disturbance	35
Article 11	Diffraction. Pulse Propagation	39
Article 12	Seismic and Vibrational Problems	43
Article 13	Concluding Notes	44
	Bibliography	45

THE MATHEMATICAL THEORY OF PLASTICITY

Introduction   51

Chapter 1   Theory of Perfectly Plastic Solids   55

     1   GENERALIZED VARIABLES   55
     2   YIELD CONDITION AND FLOW LAW   57
     3   DEFINITION OF PROBLEMS   61

Chapter 2   Theory of Strain-Hardening Plastic Solids   63

     4   YIELD CONDITION AND FLOW LAW   63
     5   KINEMATIC HARDENING   65
     6   ISOTROPIC HARDENING   71
     7   OTHER TYPES OF HARDENING   73

Chapter 3   Piecewise Linear Plasticity   75

     8   PERFECTLY PLASTIC SOLIDS   75
     9   STRAIN-HARDENING SOLIDS   77

Chapter 4   Minimum Principles of Plasticity   81

     10   INTRODUCTION   81
     11   RATE PRINCIPLES   84
     12   FINITE PRINCIPLES   89
     13   LIMIT ANALYSIS   91

Chapter 5   Bending of a Circular Plate   95

     14   RIGID-PERFECTLY PLASTIC MATERIAL   95
     15   ELASTIC-PERFECTLY PLASTIC MATERIAL   98
     16   RIGID-STRAIN HARDENING MATERIAL   102
     17   DYNAMIC LOADING   106
     18   APPLICATION OF PRINCIPLE OF MINIMUM POTEN-
           TIAL ENERGY   109

Chapter 6   Other Problems   113

     19   CIRCULAR CYLINDRICAL SHELL   113

20 PLANE STRAIN AND PLANE STRESS   117
21 BEAMS, BARS, AND RODS   119
22 MISCELLANEOUS PROBLEMS   120

Chapter 7 Russian Contributions   121

23 GENERAL REMARKS   121
24 CONTRIBUTIONS UP TO 1949   123
25 CONTRIBUTIONS FROM 1949 TO 1955   126

Bibliography   128

Author Index   145

Subject Index   149

The Mathematical Theory of
ELASTICITY

J. N. Goodier
Stanford University

THE MATHEMATICAL THEORY
OF ELASTICITY

1. Introduction. Scope and intention

An inquirer seeking a representative account of the theory of elasticity
at the present time will not find it between any one pair of covers, nor
in any one language. In 1927, the date of the last edition of Love's
treatise, he might almost have done so, and Love's interpretation of
"The Mathematical Theory of Elasticity" was wider than ours will
be, for it included plates and shells. He will find that in the past 5
years a "one-foot shelf" of new books has appeared which, in company
with the older books, provides an excellent comprehensive treatment
of the subject, and a basis for the understanding and evaluation of the
steady stream of research papers.

The proposed short list for the one-foot shelf is, in inverse chrono-
logical order: A. E. Green and W. Zerna, *Theoretical Elasticity* (1954);
L. A. Galin, *Contact Problems of the Theory of Elasticity* (in Russian)
(1953); N. I. Muskhelishvili, *Some Basic Problems of the Mathematical
Theory of Elasticity* (3rd Russian edition, 1949, translated into English by
J. R. M. Radok,* 1953; a 4th Russian edition appeared in 1954); N. I.
Muskhelishvili, *Singular Integral Equations* (2nd Russian edition, 1946,
translated into English by J. R. M. Radok, 1953); G. N. Savin, *Con-
centration of Stress around Holes* (in Russian) (1951); S. G. Lekhnitzki,
Theory of Elasticity of Anisotropic Bodies (in Russian) (1950).

The preponderance of Russian titles is striking, and these books are
based almost entirely on recent Russian investigations.† Some 250

---

* The transliteration of Russian names here follows the scheme adopted in this
translation.

† Many sections of the 2nd edition of I. S. Sokolnikoff's *Mathematical Theory of
Elasticity*, McGraw-Hill, New York, 1956, are similarly based, and contain numerous
references to Russian books and papers.

Russian papers on the subject, in the narrower sense adopted here, have been noticed in *Mathematical Reviews* and *Applied Mechanics Reviews* since 1940. Most of these, all the books of the short list above, and some others, have been available for this survey.

It is not intended to be an exhaustive and proportioned survey of all branches of the subject. The limitations of space, time, and competence would forbid that in any case. Its principal aim is to draw attention to those significant recent developments believed least known to readers whose first language is English. What is relatively well known or easily accessible in this sense is omitted or touched on only briefly and broadly. The bibliography includes only those books and papers actually discussed or cited.

2. Plane stress and plane strain in the isotropic medium

The major development of the present century in this branch of the subject has occurred chiefly in the work of Muskhelishvili and in the numerous investigations inspired by it. The importance and promise of the new methods and results were recognized and made available in English by I. S. Sokolnikoff some 15 years ago. They were concerned with the problems of prescribed boundary forces (the *first fundamental problem*) and of prescribed boundary displacements (the *second fundamental problem*). The later development has included the problem of mixed boundary conditions—the specified conditions involving both force and displacement, as when force is prescribed on one part of the boundary, displacement on the remainder—(the *third fundamental problem*). It is set forth, as of 1949, in Radok's translation of the book by Muskhelishvili [1].

While the new method for the third problem was emerging, there was extensive application of Muskhelishvili's methods for the first and second problems, predominantly in Russia. Most of this finds only mention in Muskhelishvili's own book, devoted as it is to method and comprehensive forms of solution rather than to detailed application. Here the book by Savin [1] is a valuable supplement, but it has not been translated.* It provides complete solutions and many detailed evaluations for a great variety of problems of stress concentration at holes, several of which have been worked out independently in recent non-Russian papers. The book is not, however, limited to the plane problems of the homogeneous isotropic medium as dealt with in Muskhelishvili's book. It deals as extensively with stress concentration in

---

* Since this was written a German translation by H. Neuber has been published: G. N. Sawin, *Spannungserhöhung am Rande von Lochern*, V E B Verlag Technik, Berlin, 1956.

the anisotropic medium, and also with the closely similar thin plate flexure problems, isotropic and (briefly) anisotropic. The bibliography of 135 items is naturally mainly Russian, but the relevant non-Russian literature appears to be fairly adequately represented, and some sections of the book are founded on it.

The basis for the isotropic problems is the now well-known Kolosov representation of stress $(\sigma_x, \sigma_y, \tau_{xy})$ and displacement (u, v) in terms of two complex potentials $\phi(z), \psi(z)$,

$$(1) \qquad \sigma_x + \sigma_y = 2[\phi'(z) + \overline{\phi'(z)}], \qquad z = x + iy, \qquad \bar{z} = x - iy$$

$$(2) \quad \sigma_y - \sigma_x + 2i\tau_{xy} = 2[\bar{z}\phi''(z) + \psi'(z)]$$

$$(3) \qquad 2\mu(u + iv) \;= \kappa\phi(z) - \overline{z\phi'(z)} - \overline{\psi(z)}$$

where μ is the shear modulus, ν Poisson's ratio, $\kappa = 3 - 4\nu$ for plane strain and $(3 - \nu)/(1 + \nu)$ for plane stress. These formulas for rectangular components lead very readily to components in curvilinear co-ordinates ξ, η, derived from a conformal mapping $z = \omega(\zeta), \zeta = \xi + i\eta$, which is usually to a unit circle in the ζ-plane of mathematical operation, from the physical region in the z-plane. The character of the parts of the complex potentials $\phi(z), \psi(z)$ which are nonholomorphic (in the hole or in the material region), corresponding for instance to resultant force on a hole, or dislocation discontinuity at a cut, is made out in Muskhelishvili's book. The determination of the holomorphic parts remains. He showed that this may be effected from the boundary conditions by using the well-known Cauchy integral formula for analytic functions, although the application of this formula is not immediate and direct. Theorems derived from it, apparently for this purpose by Muskhelishvili himself, are required and are given with proofs in his book [1].

Savin [1] provides detailed treatment, in most cases with tables, curves, and charts, of an extensive set of particular problems of stress concentration at holes.* The hole shapes are of four types: (1) the rectangle with rounded corners, (2) the triangle with rounded corners, (3) the ellipse, (4) the circle. These are taken with and without reinforcement, as disturbances of uniform normal or shearing stress, or simple distributions (bending, cantilever bending).

The actual shapes for which detailed results are given are selected, in types (1) and (2), by forming the Schwarz-Christoffel transformation

---

* Burmistrov [1] in a recapitulation of some of these solutions gives corrections to Savin's evaluations.

for the exact rectangle or triangle, and developing it as a series, as, for instance, for the square:

$$(4) \qquad z = \omega(\zeta) = c\left(\frac{1}{\zeta} - \tfrac{1}{6}\zeta^4 + \tfrac{1}{56}\zeta^7 - \tfrac{1}{176}\zeta^{11} + \cdots\right)$$

Then the first two, three, or four terms only are retained, to give three shapes progressively closer to the square. Problems of holes of the ovaloid forms (c, m, n constants)

$$(5, 6) \qquad z = \omega(\zeta) = c\left(\zeta + \frac{m}{\zeta} + \frac{n}{\zeta^3}\right); \qquad z = c\left(\zeta + \frac{m}{\zeta^n}\right)$$

have been solved (with evaluations) by Greenspan, Morkovin, and Green respectively.* There is of course no necessity in Savin's choice of the coefficients as those of the Schwarz-Christoffel expansion. Technical interest is likely to be focused, not on "ideal" holes with perfectly sharp corners and therefore usually infinite stress concentrations, but on corners rounded in a geometrically simple fashion, for which reliable stress estimates can be made. This question of choice of shape for evaluations such as those given by Savin is one of some importance, and will be raised again later.

The book gives a brief approximate treatment of a problem of considerable engineering interest which, as far as the writer has observed, has not yet appeared in the non-Russian literature. For a circular hole in a thin cylindrical shell, the effects of the curvature of the shell are represented in the stress-concentration factors (on longitudinal stress)

$$(7, 8) \qquad k_1 = 3\left(1 + 0.43\,\frac{\rho_0^2}{ah}\right), \qquad k_2 = 2.5\left(1 + 2.3\,\frac{\rho_0^2}{ah}\right)$$

for fields of longitudinal tension, and of longitudinal and circumferential tensions as in a pressure cylinder with closed ends. Here ρ_0 is the radius of the hole, a the radius, and h the thickness of the shell. The source is Lourie [1, 2].

Savin's accounts of the strip with a centrally placed circular hole, the circular hole in the semi-infinite plate, and two holes intersecting or nonintersecting, are in terms of Fourier methods and real variables, and in the main are drawn from the work of Howland, Mindlin, and Ling.†

---

* See, for instance, the references on p. 212 of Timoshenko and Goodier, *Theory of Elasticity*, McGraw-Hill, New York, 1951. Also see Green and Zerna [1], p. 296.

† See, for instance, the references given in Timoshenko and Goodier, *op. cit.*, p. 211.

A recent addition to calculated results of this kind is given by Isida [1], for a strip with an eccentric circular hole, under bending and under tension, and is complemented by photoelastic confirmation. References of the period 1940–1950 may be found in an earlier survey.* A paper on the strip with a semicircular notch in each edge by Ling [1] constructs a sequence of stress functions each member of which is an infinite series (here of Fourier type). Superposition of such members, or their construction by satisfying first one boundary, then another, is of course well known, and is exemplified by numerous papers on combinations of circular and straight boundaries (as Howland's †). But Ling effects an improvement in these sequences, under the term "promotion of rank." The earlier Fourier harmonics of each member of the sequence are removed by suitable subtraction of preceding members, and in the new sequence each member starts at a Fourier harmonic of higher order than its predecessor.

In general, problems of regions of more than twofold connectivity have not yielded to the complex variable methods, since they are not amenable to conformal transformation within or outside the unit circle. Muskhelishvili [1] (Radok's translation, p. 397) touches on reductions to Fredholm integral equations by Sherman and Mikhlin. In later papers Sherman [1, 2, 3] devotes lengthy analyses in the complex variable to the heavy vertical plate (putting the matter in plane stress terms) with two circular holes, or with two elliptical holes. A free straight horizontal upper edge is contemplated, but the problems really concern the infinite plate, the holes being taken too small and too far down to be influenced by the free edge. The circular hole cases seem suitable for the general Fourier form of solution in bipolar co-ordinates constructed by Jeffery in 1921, and used for numerous particular problems since. On the whole it appears that the multiply-connected regions stand in the way of a general unification of the two-dimensional theory in terms of complex function theory.

Gravitational stress affected by noncircular holes or tunnels is not considered in Savin's book. Yu [1] has derived expressions for the complex potentials and the stress at the hole for the ovaloid holes of the transformation (5). He has also given the forms taken in a Muskhelishvili representation by the solutions for the heavy circular annulus supported at a point, and the heavy disk supported by concentrated forces. Savin does include the annulus with diametrically opposite forces (the "proving-ring problem") and gives some Russian evaluations.

* Goodier, *Appl. Mechanics Revs.*, **4**, 1951, p. 330.

† See the references given in Timoshenko and Goodier, *op. cit.*, p. 83.

3. Holes and fillets of assignable shapes.
Approximate conformal mapping

The ovaloid holes given by the transformations (4), (5), and (6) have been extensively studied and their stress concentrations evaluated, because they are mathematically convenient and give shapes approaching rectangles, rhombi, and triangles with rounded corners. The practical estimation of such studies will depend on whether they can be used with confidence for actual holes, which commonly will have straight sides and circular arc roundings at the corners. The shape constants m, n of the transformation (5) can be adjusted to give, for a rectangle, the required ratio of sides, and the ratio of the minimum radius of curvature at the corners to a side. But the rounding off is not of circular arc form, and the sides are not quite straight. In general, the attraction of mathematically convenient exact conformal mappings tends to make the development of the two-dimensional theory resemble that of the Saint-Venant theory of torsion, in which a large accumulation of results is available for mathematically simple but otherwise very improbable shapes.

Greater practical interest would be engaged by a method which could cope directly with an actual given shape. Such a method is offered by Kikukawa [1, 2, 3], and its possibilities demonstrated by complete calculations for a number of striking examples of stress concentration in the stretched plate with (a) a hole of rhombic form, with circular arc fillets at the corners; (b) a double notch, each notch having straight parallel sides joined to U-shape by a semicircle; (c) circular quadrant fillets at a change from finite to infinite width. The results for case (b) agree very well with those of Neuber for the double hyperbolic notch. The results for case (a) show, if they are valid, that the ovaloid approximation to the shape attainable by a mapping such as (5) is *not* adequate. No previous theoretical evaluation of case (c) is known to the writer, although there are photoelastic measurements.

Kikukawa's method proceeds from an initial mapping $z = z_0(\zeta)$ in simple definite form, for instance, the mapping (5) for case (a) above. This is amended to bring the hole shape sufficiently close to the prescribed form. The method of Theodorsen and Garrick, devised for airfoil shapes, is not found suitable.*

The initial mapping $z = z_0(\zeta)$ gives a curve C_0 (e.g., ovaloid) in the z-plane somewhat different from the given curve C [e.g., the hole of case (a) above], as shown in Fig. 1b. The point Q_0 on C_0 corresponds

* A survey of numerical methods in conformal mapping is given by G. Birkhoff, D. M. Young, and H. Zarantonello, in *Proc. Symp. Appl. Math.*, **4**, 117, 1953.

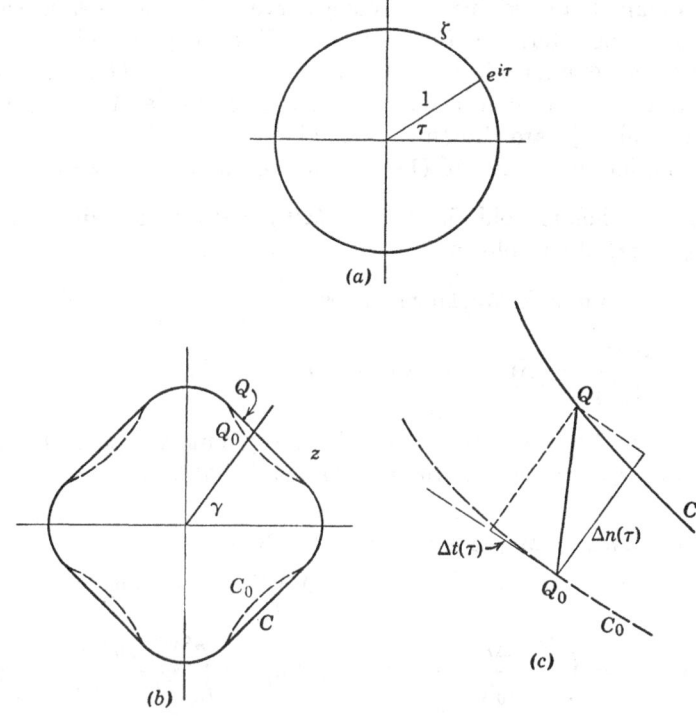

Fig. 1.   Kikukawa's approximate conformal mapping.

to a point $\zeta = e^{i\tau}$ on the unit circle of the ζ-plane (Fig. 1a). The point Q is the unknown, but neighboring, point on C derived from the desired, but of course unknown, mapping function $z(\zeta)$. Writing

$$z(\zeta) - z_0(\zeta) = \Delta z(\zeta)$$

the vector $Q_0 Q$ is $\Delta z(\zeta)$ at $\zeta = e^{i\tau}$. If $\Delta n(\tau)$ and $\Delta t(\tau)$ are the components of $Q_0 Q$ along the normal and tangent to C_0 at Q_0, we have (Fig. 1c)

(9) $$\Delta n(\tau) + i\,\Delta t(\tau) = \Delta z(e^{i\tau}) \cdot e^{-i\gamma}$$

where γ, the angle between the normal and the x-axis, is given by

(10) $$e^{i\gamma} = \zeta \cdot z_0{}'(\zeta)/|z_0{}'(\zeta)| \qquad \text{with } \zeta = e^{i\tau}$$

From (9) and (10)

(11) $$\frac{\Delta z(\zeta)}{\zeta \cdot z_0{}'(\zeta)} = \frac{\Delta n(\tau) + i\,\Delta t(\tau)}{|z_0{}'(\zeta)|} \qquad \text{with } \zeta = e^{i\tau}$$

The domain bounded by the single curve C in the z-plane can be mapped into the interior of the unit circle T in the ζ-plane by $z = z(\zeta)$, where $z(\zeta)$ is holomorphic in T and $z'(\zeta) \neq 0$ in T. The mapping is unique under the normalization conditions $z(0) = a$, and $\arg z'(0) = 0^{\cdot}$ Both $z(\zeta)$ and $z_0(\zeta)$ are chosen accordingly.

The function on the left of (11) is holomorphic in T because

(i) $\Delta z(\zeta)$ is holomorphic in T, and $\Delta z(0) = 0$ by normalization.

(ii) $z_0'(\zeta)$ is holomorphic in T.

There is accordingly a Maclaurin series

$$(12) \qquad \frac{\Delta z(\zeta)}{\zeta \cdot z_0'(\zeta)} = \Delta m(\zeta) = \Delta m_0 + \Delta m_1 \cdot \zeta + \Delta m_2 \cdot \zeta^2 + \cdots$$

where $\Delta m_p = \Delta m_p' + i\,\Delta m_p''$ $(p = 1, 2, \cdots)$, but Δm_0 is real because $\Delta z'(0)$ and $z_0'(0)$ are real. It follows from (11) that

$$\frac{\Delta n(\tau)}{|z_0'(e^{i\tau})|} = \Delta m_0 + \Delta m_1' \cos \tau + \Delta m_2' \cos 2\tau + \cdots$$
$$- \Delta m_1'' \sin \tau - \Delta m_2'' \sin 2\tau - \cdots$$

and so

$$(13) \quad 2\pi\,\Delta m_0 = \int_0^{2\pi} \frac{\Delta n(\tau)}{|z_0'(e^{i\tau})|}\,d\tau; \qquad \pi\,\Delta m_p' = \int_0^{2\pi} \frac{\Delta n(\tau)}{|z_0'(e^{i\tau})|}\cos p\tau\,d\tau$$

$$\pi\,\Delta m_p'' = \int_0^{2\pi} \frac{\Delta n(\tau)}{|z_0'(e^{i\tau})|}\sin p\tau\,d\tau$$

Now if $\Delta n(\tau)$ were known, the coefficients of the series (12) could be evaluated by these formulas (13). In fact, we have only the two curves C_0 and C, and do not know the point Q which, by the desired mapping, corresponds to $e^{i\tau}$ on the unit circle $|\zeta| = 1$. But $\Delta n(\tau)$, being the projection of Q_0Q on the normal to C_0 at Q_0, is nearly equal to the distance $\delta n(\tau)$ from C_0 to C along the normal. This is known, and it is used in place of $\Delta n(\tau)$ to evaluate approximately the coefficients (13) of the series (12). The complex potentials of (1), (2), and (3) are found by a simple expansion procedure.

For the hole of case (a) above, Kikukawa finds that two terms of (12) make a substantial difference to the stress calculated for the initial ovaloid, but the third term brings little further change. If these results are confirmed, the method will add much to the effectiveness of the two-dimensional theory, as well as to plate flexure theory. These biharmonic problems are much less suitable for numerical methods, as relaxation, than are the potential problems of torsion or flexure.

4. Reinforcement of holes

The artificiality of the (in the exact sense) mathematically amenable forms is no less troublesome in the problem of the hole reinforced by a ring. The circular hole and ring offer no difficulty, and have been well investigated. References may be found in the recent paper by Heller [1] and in the writer's earlier survey (footnote, p. 7). Problems of nests of circular rings in holes (a hole with a ring welded inside it, then another, and another, etc.), the rings being of different materials, appear in several Russian papers. The usual fields of tension, bending, and shear prevail at infinity. Each ring as well as the plate is treated by plane stress theory. References, and evaluations for a single ring of different material (or equivalently under plane stress assumptions, of a different thickness), are given in Ch. 5 of Savin [1]. Being for specialized values, the evaluations do not cover a representative range of conditions, and are merely illustrative. The other investigations have proceeded independently, most of them treating the ring as a curved bar (the "bead" reinforcement). Heller [1], in an analysis of this kind for a reinforced circular hole centrally placed in a plate carrying bending and shear stress, finds as Reissner and Morduchov [1] had suggested, that the reinforcement does its work in tension or compression rather than bending.

When the hole is not circular, and both ring and plate are treated by the plane stress theory, there is at once a mathematical difficulty as to the shape of the ring. This shape is specified by the inner free boundary curve, and the outer ring boundary curve where the ring is attached to the plate. The analysis is manageable only if these two curves belong to the same family—e.g., they map into concentric circles in the ζ-plane. For the simplest case—the elliptical hole—the two bounding curves become confocal ellipses. For ovaloid shapes they are similarly restricted. The example in Fig. 2, reproducing one of Savin's diagrams (his Fig. 126), is striking in that the smallest ring thickness (in the plane) appears at the corners where the stress concentration due to a hole is greatest—a questionable feature from a practical point of view.

Problems of this kind have been worked out by Seremetev [1] for the mapping (6) (this mapping is discussed on p. 177 of Muskhelishvili [1], Radok's translation, and on p. 296 of Green and Zerna [1]), independently by Wells [1] for the elliptic case, and by Harvey [1] in general terms for rings of small thickness (in the plane) and specifically for the ovaloid mapping (5). Savin [1] gives an account of Seremetev's investigations with extensive evaluations of (for each shape) a particular illustrative elastic ring in comparison with the absolutely rigid ring and the unreinforced hole. Figure 2 is an example. Figure 3 from Savin [1] (Fig. 132, p. 344) shows the variety of shapes examined, with results for maximum stress in each case.

The results for the absolutely rigid ring (the rigid inclusion) are of course dependent only on the shape of the outer bounding curve of

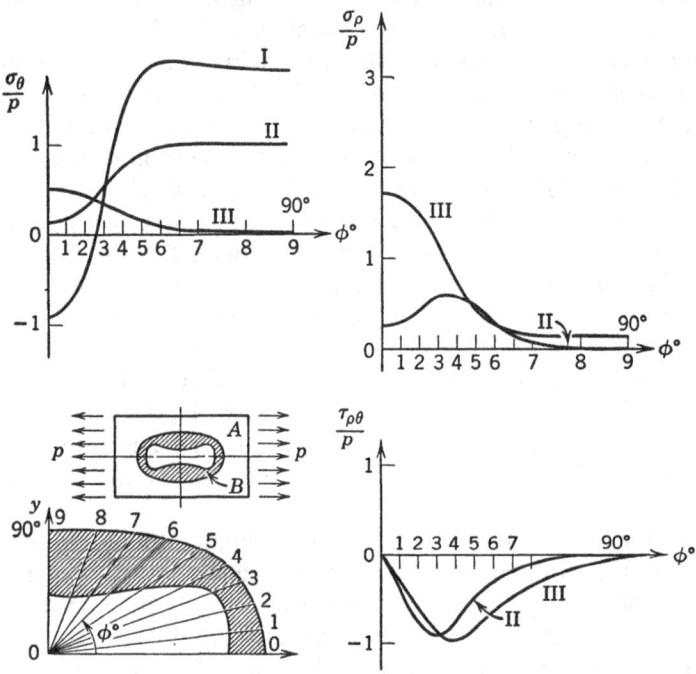

Fig. 2. Ovaloid hole with and without reinforcement, in a field of tension. Stresses in plate at ring, or hole. Circumferential stress σ_θ, normal stress (zero for hole) σ_ρ, shear stress $\tau_{\rho\theta}$. Curve I is for the hole (no reinforcing ring); curves II are for a steel ring in a copper (elastic) plate; curves III are for a rigid ring in an elastic plate. (Figure 126 of G. N. Savin, *Concentration of Stress around Holes.*)

the ring, not on the inner boundary. It seems very desirable now that these problems, especially in the case of the rigid inclusion, should be worked out by the method of Kikukawa (Section 3 above) for "rectangles," "rhombi," "triangles," and slot shapes formed of straight sides and circular arcs.

Treatment of the reinforcement as a thin curved bar releases the analysis from the difficulty as to ring shape pointed out above, and also from the plane stress theory restriction to thin rectangular cross section. It tends to a general simplification, accepting the elementary thin rod equations for the ring. These at once provide boundary conditions for the plate at the hole, and the Kolosov-Muskhelishvili methods can be applied. This type of analysis for the reinforced noncircular hole is carried out in general terms in a later paper by Seremetev [2], who includes the flexural stiffness of the ring, and refinements such as change of curvature due to extension of center line. Detailed application is to the circular hole only, with a uniform ring, and it is suggested

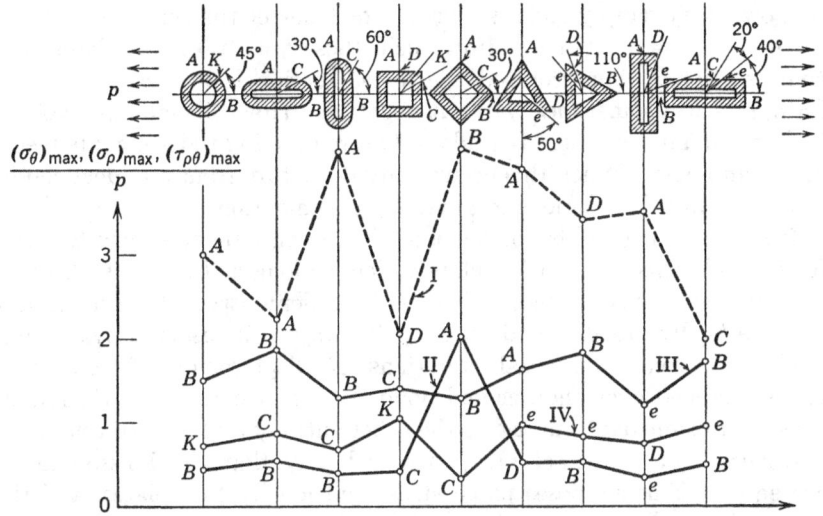

Fig. 3. Stress at holes and at rigid inclusions of various shapes. Significant values at individual points A, B, C, etc. I—circumferential stress (σ_θ) at hole (no reinforcing ring). II—circumferential stress (σ_θ) at rigid ring. III—normal stress (σ_ρ) at rigid ring. IV—shear stress ($\tau_{\rho\theta}$) at rigid ring. (Figure 132 of G. N. Savin, *Concentration of Stress around Holes.*)

that not only the refinements but also the flexural stiffness can usually be disregarded. The similar analysis required for the anisotropic plate is given in detail, but limited to the circular reinforced hole. The field of stress is throughout uniform at infinity. Seremetev [3] has since examined the plate flexure problem also.

In an independent treatment of the reinforcement problem on the same basis, Radok [1] provides a concise formulation of the general (plane stress, isotropic) equations by discarding flexural stiffness and refinements to begin with. Here also evaluation is limited to the circular hole and the uniform ring. The boundary condition imposed by the ring on the complex potentials of the plate is found, and serves to determine the coefficients in their Laurent expansions in negative powers of z. It appears possible that Kikukawa's method holds promise of extracting from this theory results for assigned practical noncircular shapes of hole.

The concern of structural design is primarily of course to eliminate the stress disturbance at the reinforced hole, and only secondarily to evaluate it when present. The possibility of complete elimination by a flexible ring is put forward in a paper by Mansfield [1]. In an undisturbed stress field represented by an Airy stress function $\phi(x, y)$, it is shown very simply to be realizable for holes of the shapes $\phi(x, y) + ax + by + c = 0$. For a uniform field these "neutral hole" shapes are conic sections. The required ring cross section is in general nonuniform, and will sometimes turn out negative. How this may be avoided is shown in an example. It leads to two reinforcing rod segments meeting at an angle. When this occurs, any departure from the ideal condition may introduce serious local stress concentration.

If a reinforcing rod is not a closed ring, it may be necessary to allow for forces between it and the plate which, so long as the rod is idealized as a line, are concentrated. This has been demonstrated, with support from tests, by Goodier and Hsu [1], by way of a result derivable immediately from differential equations of displacement: if one thin plate is lapped over another and bonded to it over the area of contact, force is transmitted from one to the other entirely by line load round the periphery of the contact area, provided only that the Poisson ratios are equal. This supposes plane stress and ideally thin plates, and the line load actually presents of course a local stress concentration problem. It indicates that in investigations of tension communicated to a plate by an attached bar (in the plane), as in a recent paper * of Koiter [1], forms of solution which by their regularity postulates exclude concentrated forces may not be admissible.

5. Mixed boundary value problems. The third fundamental problem in two dimensions

While the earlier papers of Muskhelishvili systematized the first and second fundamental problems (for suitably transformable regions) in an elegant relation to complex function theory, it is the later contributions on mixed boundary conditions—the third fundamental problem—which have added most substantially to the list of new solutions. This development appears to have no counterpart outside Russia. It is, fortunately, included in Radok's translation of Muskhelishvili [1]. The mathematical basis—complex singular integral equations—with several of its applications in other branches of applied mathematics, is

* References to previous papers on this subject may be found in Goodier and Hsu [1].

set forth in another book by Muskhelishvili [2], also translated by Radok.

The method may be briefly indicated by an example: the circular disk with displacements prescribed on n assignable segments of the boundary circle (denoted collectively by L'), the intervening segments (denoted collectively by L'') being free (Muskhelishvili [1], p. 510). The method is not confined to the circle. Regions which can be mapped in a circle by rational functions are dealt with in Ch. 21 of Muskhelishvili [1].

The formulas (1), (2), and (3) for stress and displacement in terms of two complex potentials $\phi(z)$, $\psi(z)$ are readily converted to curvilinear co-ordinates derived from $z = \omega(\zeta)$. In the special case needed here—polar co-ordinates ($z = re^{i\theta}$)—they become [writing $\Phi(z) = \phi'(z)$, $\Psi(z) = \psi'(z)$],

(14) $$\sigma_r + \sigma_\theta = 2[\Phi(z) + \overline{\Phi(z)}]$$

(15) $$\sigma_\theta - \sigma_r - 2i\tau_{r\theta} = 2e^{2i\theta}[\bar{z}\Phi'(z) + \Psi(z)]$$

and from these follows

(16) $$\sigma_r - i\tau_{r\theta} = \Phi(z) + \overline{\Phi(z)} - e^{2i\theta}[\bar{z}\Phi'(z) + \Psi(z)]$$

For the displacements the rectangular components u, v are retained and

(17) $$2\mu(u + iv) = \kappa\phi(z) - z\overline{\phi'(z)} - \overline{\psi(z)} + \text{constant}$$

On the boundary circle $z = t = e^{i\theta}$ the conditions are

(18) $\sigma_r{}^+ - i\tau_{r\theta}{}^+ = 0$ \qquad on the free segments L''

(19) $u^+ + iv^+ = g(t)$ \qquad given on the remaining segments L'

The superscripts $+$ denote limits as z approaches a boundary point t from S^+, the interior of the circle.

The method consists in first satisfying (18) by devising a continuation of $\Phi(z)$ through L'' to the region S^- exterior to the circle. Then $\Psi(z)$ becomes determinable from $\Phi(z)$, and $\Phi(z)$ itself is found from (19).

Corresponding to a function $\Phi(z)$ there is another defined by $\bar{\Phi}(z) = \overline{\Phi(\bar{z})}$, bars indicating complex conjugates. Like $\Phi(z)$, it is analytic in S^+; its Maclaurin series comes from that of $\Phi(z)$ by changing the coefficients to their conjugates. As yet $\Phi(z)$ is defined only in S^+. The continuation through L'' is devised as

(20) (for z in S^-) $$\Phi(z) = -\bar{\Phi}\left(\frac{1}{z}\right) + \frac{1}{z}\bar{\Phi}'\left(\frac{1}{z}\right) + \frac{1}{z^2}\bar{\Psi}\left(\frac{1}{z}\right)$$

Examination of the right-hand side reveals a function defined and holomorphic in S^-, since when z is in S^-, $1/z$ is a point in S^+.

On the other hand, when z is in S^+, $1/\bar{z}$ is a point in S^- and can be used in (20). Then

$$(21) \qquad \text{(for } z \text{ in } S^+) \qquad \Phi\left(\frac{1}{\bar{z}}\right) = -\overline{\Phi}(\bar{z}) + \bar{z}\overline{\Phi}'(\bar{z}) + \bar{z}^2\overline{\Psi}(\bar{z})$$

This yields $\overline{\Psi}(\bar{z})$, and hence its complex conjugate $\Psi(z)$ for z in S^+, in terms of $\Phi(z)$ and $\Phi'(z)$ as continued by (20). The formulas (1), (2), (3), and (4) can now be expressed in terms of $\Phi(z)$ defined for the whole plane except on the circle.

In (16), $e^{2i\theta}$ can be replaced by $z/\bar{z}$, and conjugates taken throughout. It becomes

$$\sigma_r + i\tau_{r\theta} = \Phi(z) + \overline{\Phi(z)} - \bar{z}\overline{\Phi'(z)} - \frac{\bar{z}}{z}\overline{\Psi(z)}$$

and when (21) is used to remove the term $\bar{z}\overline{\Phi'(z)}$

$$(22) \qquad \sigma_r + i\tau_{r\theta} = \Phi(z) - \Phi\left(\frac{1}{\bar{z}}\right) + \bar{z}\left(\bar{z} - \frac{1}{z}\right)\overline{\Psi(z)}$$

This is a convenient form for use in the boundary condition (18). If z (in S^+) approaches a point t on the boundary circle, $1/\bar{z}$ (in S^-) also approaches t, and $\bar{z} - 1/z$ approaches zero. The last term in (22) disappears in the limit. $\Phi(z)$ has some limit $\Phi^+(t)$ from S^+, and $\Phi(1/\bar{z})$ some limit $\Phi^-(t)$ from S^-, and (22) becomes

$$(23) \qquad \lim_{z \to t} (\sigma_r + i\tau_{r\theta}) = \Phi^+(t) - \Phi^-(t)$$

On the free segments L'', $\sigma_r + i\tau_{r\theta}$ vanishes, and so (23) yields

$$(24) \qquad \Phi^+(t) - \Phi^-(t) = 0 \qquad \text{on } L''$$

Consequently $\Phi(z)$, as continued by (20), preserves continuity through L'', and is holomorphic in the whole plane except on L', where $\sigma_r + i\tau_{r\theta}$ will not vanish.

The remaining boundary condition (19) now provides a relation between the inside and outside limits $\Phi^+(t)$ and $\Phi^-(t)$ on L'. Writing u', v' for $\partial u/\partial\theta$, $\partial v/\partial\theta$, (19) yields by differentiation with respect to θ

$$(25) \qquad u'^+ + iv'^+ = g'(t) \qquad \text{on } L'$$

and (17) yields generally, after using (21) to remove $\bar{z}\overline{\Phi'(z)}$

$$(26) \qquad 2\mu(u + iv) = iz\left[\kappa\Phi(z) + \Phi\left(\frac{1}{\bar{z}}\right) - \bar{z}\left(\bar{z} - \frac{1}{z}\right)\overline{\Psi(z)}\right]$$

With this (25) becomes

(27) $$\kappa\Phi^+(t) + \Phi^-(t) = 2\mu g'(t) \qquad \text{on } L'$$

and this is the governing condition for the function $\Phi(z)$ holomorphic except on L'.

It presents a special case of the *problem of linear relationship* (*the Hilbert problem*). In the general case the constant κ is replaced by a given function of t, and the discrete circular arcs L' are replaced by discrete arcs of general smooth nonintersecting form. The complete solution of this problem is a recent development in the theory of complex functions. It is given, with its history and precise formulation, in Muskhelishvili [2] (Ch. 5).

The solution, and with it the new method for mixed boundary value problems, rests on a theorem of Plemelj concerning Cauchy integrals of the type

(28) $$F(z) = \frac{1}{2\pi i} \int_{L'} \frac{f(t)\, dt}{t - z}$$

—that the limits $F^+(t_0)$, $F^-(t_0)$ from the two sides of L', at a point t_0 of L', are given by

(29) $$F^+(t_0) - F^-(t_0) = f(t_0)$$

(30) $$F^+(t_0) + F^-(t_0) = \frac{1}{\pi i} \int_{L'} \frac{f(t)\, dt}{t - t_0}$$

under precise conditions which will not be stated here. This theorem is cited and used without proof in Muskhelishvili's *Some Basic Problems of the Mathematical Theory of Elasticity* [1], and also in Green and Zerna [1], who make extensive use of the Hilbert problem in the two-dimensional theory. Most readers will therefore find Muskhelishvili's *Singular Integral Equations* [2] a necessary supplement to these books.

The solution of the Hilbert problem represented by (27), and with it the solution of the mixed boundary value problem of the circular disk represented by (18) and (19), is (Muskhelishvili [1], Ch. 18)

(31) $$\Phi(z) = \frac{\mu X_0(z)}{\pi i \kappa} \int_{L'} \frac{g'(t)\, dt}{X_0^+(t) \cdot (t - z)} + X_0(z) P_n(z)$$

where

$$X_0(z) = \prod_{k=1}^{n} (z - a_k)^{-\frac{1}{2} - i\beta} (z - b_k)^{-\frac{1}{2} + i\beta},$$ meaning the branch which for large $|z|$ has the form $z^{-n} + \alpha_{-n+1} z^{-n+1} + \cdots$;

a_k and b_k are the end points of the kth of n arc segments constituting L';
$\beta = \frac{1}{2}\pi \log \kappa$, κ being an elastic constant;
$P_n(z) = C_0 z^n + C_1 z^{n-1} + \cdots + C_n$, where $C_0 \cdots C_n$ are constants determined by satisfying the undifferentiated boundary condition (19).

Muskhelishvili [1, 2] shows that integrals such as that in (31) can be evaluated when $g'(t)$ is a polynomial or a rational function. His *Some Basic Problems* contains and implies a wealth of results for mixed problems of the disk, the circular hole, the elliptic hole, the semi-infinite plate (or solid in plane strain) with edge indented by a rigid stamp,* with and without adhesion and friction, several such stamps separate or linked, the infinite plate with circular arc slits, and for certain cases of the smooth rigid oversize disk inserted in a hole in the infinite, elastic plate (or elastic disk in rigid plate). The oversize is a given function of position on the hole, and all-round contact is assumed. This is a mixed problem since the normal displacement (the oversize) and the tangential force (zero) are prescribed, and the tangential displacement and normal contact force remain to be found. The result is given for (elastic) regions mapped in the circle by rational functions. The illustrative examples given in detail are for the circular disk, the circular hole, and the elliptic hole. Since all-round contact is postulated, these cases appear suitable for elementary Fourier treatment.

It is in problems where all-round contact does not occur that the method shows its power to best effect. For instance, if a rigid or elastic disk just fits into a smooth circular hole in the infinite elastic plate, uniform principal stresses σ_1, σ_2 at infinity may result in contact over only part, or even none, of the circle. The corresponding ranges of σ_1 and σ_2 can of course be found from the elementary solutions for the circular hole. When the contact is only partial, the arcs of contact have to be found in the course of the solution. This problem is worked out in a paper of Seremetev [4] subsequent to Muskhelishvili [1].

Another novel type of problem is represented by a paper of Mossakowskii and Zagubizenko [1]. An infinite plate has a straight slit in it, of small uniform width δ. A single uniform principal stress $-p$ at infinity acts at an arbitrary angle to the slit. Under sufficient pressure p the slit deforms and closes on some central length, which of course is not known in advance. Following Muskhelishvili's treatment of the first fundamental problem for the half plane (Muskhelishvili [1], p. 496), the solution is obtained in integrals of the type appearing in (31), L'

---

* A solution for this problem, on a different basis, due to H. G. Hopkins, is given in I. N. Sneddon's *Fourier Transforms*, McGraw-Hill, New York, 1951, p. 431.

now consisting of straight segments. These are evaluated to give an equation for α, the half length of the closure, in the form

$$2E\left(\frac{\pi}{2}, \sqrt{1 - \frac{\alpha^2}{a^2}}\right) - \frac{\alpha}{a} F\left(\frac{\pi}{2}, \sqrt{1 - \frac{\alpha^2}{a^2}}\right) = \frac{2\mu\delta}{pa(\kappa + 1)\sin^2\beta}$$

where E and F are elliptic integrals of the first and second kinds, a is the half length of the slit, and β is the angle p makes with the slit. When α is found, the normal contact pressure (there is no friction) is given by

$$p(t_0) = -p\sin^2\beta \sqrt{\frac{\alpha^2 - t_0^2}{a^2 - t_0^2}}$$

where t_0 is the co-ordinate of a point in the contact region $-\alpha < t_0 < \alpha$.

The term "contact problem" in the Russian literature covers indentation by rigid dies, or stamps, where the boundary of the contact zone and the form of the slightly curved indenting surface are prescribed—as well as the Hertzian problem of local contact of two elastic bodies, with the contact zone to be determined. Hertz took the undeformed boundaries in the neighborhood of the contact zone as sufficiently well represented (in the general three-dimensional case) by $z = ax^2 + by^2$ (x, y in the tangent plane) on grounds of smoothness. The Russian work has gone beyond this in both two and three dimensions. In the plane problem solutions are available (e.g., in Muskhelishvili [1]) for general forms $y = f(x)$ of bounding curve. The generalization is significant when the two surfaces very nearly mate over a considerable extent before contact, or when they have been shaped to a form not of the second degree even locally.

The variety of rigid die solutions given in Muskhelishvili's book need not be detailed here in view of the general availability of Radok's translation. Two books titled *Contact Problems of the Theory of Elasticity*, one by Galin [1] and one by Shtaerman [2], have not been translated. The contents of the former, as the later and more comprehensive treatment, will be indicated insofar as they are not covered by Muskhelishvili.

Half of Galin's book is devoted to the plane problem (the three-dimensional part will be considered in Section 10). It includes detailed results for a flat rigid die, loaded by normal force and couple (in the plane), indenting the edge of the elastic half plane. The couple causes tilting to a degree sufficient to cause separation of the deformed edge from the die at the high side, and the extent of contact is therefore to be determined. The *anisotropic* half-plane problem is solved for dies of assignable form with (unidirectional) Coulomb friction throughout the contact, and for prescribed displacement (both components) on prescribed "contact" zones with the edge otherwise free. Zones of adhesion and slip (with Coulomb friction) are dealt with in a solution for the isotropic half plane with a flat-based, level die. There is no indication of a connection with experiment. The deformation wave in the half plane, accompanying a traveling die, is discussed in Section 10.

Galin's analysis differs from Muskhelishvili's in being based not on the Hilbert problem (Section 5) but on the Riemann-Hilbert problem (cf. Muskhelishvili [2]): to find the function $\Phi(z) = u + iv$ ($z = x + iy$), holomorphic in the half plane $y > 0$, bounded at infinity, and satisfying on $y = 0$ the boundary condition,

$$a(x) \cdot u + b(x) \cdot v = f(x)$$

Here $a(x)$, $b(x)$, $f(x)$ are given real functions, usually discontinuous. His treatment of this problem differs from Muskhelishvili's ([2], Art. 43, p. 109) and he shows how the two are related.

Shtaerman's book contains a second solution of the two-dimensional contact problem for a cylinder (circular) in a bore of only slightly larger radius, involving an extended contact region. The first solution appeared in 1940 (Shtaerman [1]).

6. Eigensolutions for plane and axisymmetric states

Self-equilibrating load on the end of a semi-infinite strip can be examined through an Airy stress function (for the plane problem) which first makes the edges free, for instance

$$\phi = e^{-\gamma x}(\kappa \cos \gamma y + \gamma y \sin \gamma y)$$

with γ one of the (complex) roots of

$$(32) \qquad\qquad \sin 2\gamma + 2\gamma = 0$$

and κ an associated constant. References to papers in English and German are given by Horvay [1], who develops, for computational advantages, variational approximations in real terms.* The analogous eigenfunctions for infinite wedge regions were employed by Williams [1] for the investigation of the singularities at the tips of such regions with homogeneous boundary conditions on the two boundary lines.

The Russian references appear to begin with the book *Theory of Elasticity* by P. F. Papkovicz in 1939. Kitover [1] gives a systematic listing

---

* Applications to thermal stress problems are made by Horvay [2], and Born and Horvay [1].

of the functions and their characteristic equations [as (32) above] for strips, wedge regions, and regions bounded by radii and concentric arcs, for plane stress, and for plate flexure. The list extends further to eigensolutions for axisymmetric states in circular cylinders, drawn from Prokopov [1]. The characteristic equations are covered by the general form

$$u_0 + u_1 \sin z + u_2 \cos z = 0$$

where u_0, u_1, u_2 are polynomials of degree 0, 1, 2 in z, and roots are determined by a process which does not appear in the non-Russian papers. For instance, for $\sin z = az$, and $z = p + iq$, the introduction of

$$\delta = \sqrt{1 - \left(\frac{aq}{\cosh q}\right)^2}$$

leads to

$$\sin\left(\frac{\delta}{a}\cosh q\right) = \pm\delta$$

from which values of q can be found. Application of the eigenfunctions is illustrated by the sector of a circle clamped on the curved edge and loaded by a couple (in the plane) at the tip. Prokopov [2] has also given an analysis for a rectangular plate clamped on two opposite edges, and carrying a concentrated normal (in plane) load on one of the remaining edges. Lourie [3] has given tables and curves for a set of eigensolutions for the circular cylinder in axisymmetric deformation, referring to the problem of a band of pressure on its surface. Evaluations for this were effected otherwise by Barton and Rankin,* and for the hollow cylinder by Shapiro [1].

7. Anisotropic elasticity

An extensive development of the theory for anisotropic materials has taken place in the past 25 years in Russia, especially in the work of S. G. Lekhnitzki [1, 2], who has given a comprehensive account in two books, *Anisotropic Plates* and *Theory of Elasticity of Anisotropic Bodies*. These are supplemented by a general survey by Fridman [1], with a Russian bibliography of 88 items. The book by Savin [1] includes a chapter on elliptical (and circular) holes in anisotropic plates, providing complete solutions of the first and second fundamental problems, and numerous evaluations and curves. This also provides a supplement to Lekhnitzki's sections on these matters.

* See Timoshenko and Goodier, *op. cit.*, p. 388.

An almost parallel development began later in England in the work of Green and Taylor and was continued by Green. References may be found in the book by Green and Zerna [1]. The chapter on "Plane problems for aeolotropic bodies" provides a significant and readily accessible contribution to this branch of the subject. It exhibits the present range of the two-dimensional theory in a treatment of the basic problems by a combination of the complex representation with the Hilbert problem (cf. Section 5 above).

The complex representation arises from the Airy stress function $U(x, y)$, which meets the conditions of compatibility and the anisotropic stress-strain relations by satisfying a linear partial-differential equation of the fourth order with constant coefficients. This equation reduces to the biharmonic equation for the coefficients appropriate to the isotropic medium. The function $F(x + sy)$ is a solution if s is a root of

$$a_{11}s^4 - 2a_{16}s^3 + (2a_{12} + a_{66})s^2 - 2a_{26}s + a_{22} = 0$$

and the general form for $U(x, y)$ is

$$U(x, y) = F_1(x + s_1 y) + F_2(x + s_2 y) + F_3(x + s_3 y) + F_4(x + s_4 y)$$

with

$$\left.\begin{array}{c} s_1 \\ s_3 \end{array}\right\} = \alpha_1 \pm i\beta_1; \qquad \left.\begin{array}{c} s_2 \\ s_4 \end{array}\right\} = \alpha_2 \pm i\beta_2$$

The theory has accordingly a natural development in the two auxiliary planes of the two complex variables

(33) $$z_1 = x + s_1 y = (x + \alpha_1 y) + i(\beta_1 y)$$

(34) $$z_2 = x + s_2 y = (x + \alpha_2 y) + i(\beta_2 y)$$

and for a real U we have

$$U(x, y) = F_1(z_1) + F_2(z_2) + \overline{F_1(z_1)} + \overline{F_2(z_2)}$$

In terms of

$$\phi(z_1) = \frac{dF_1}{dz_1} = F_1'(z_1); \qquad \psi(z_2) = \frac{dF_2}{dz_2} = F_2'(z_2)$$

the stress and displacement formulas are

(35) $$\sigma_x = 2\,\mathrm{Re}\,[s_1^2\phi'(z_1) + s_2^2\psi'(z_2)]; \qquad \sigma_y = 2\,\mathrm{Re}\,[\phi'(z_1) + \psi'(z_2)]$$

(36) $$\tau_{xy} = -2\,\mathrm{Re}\,[s_1\phi'(z_1) + s_2\psi'(z_2)]$$

(37) $$u = 2\,\mathrm{Re}\,[p_1\phi(z_1) + p_2\psi(z_2)]; \qquad v = 2\,\mathrm{Re}\,[q_1\phi(z_1) + q_2\psi(z_2)]$$

The elastic constants p_1, p_2, q_1, q_2 are combinations of the elastic constants a_{11}, etc., appearing in the stress-strain relations, and the roots s_1, s_2, s_3, s_4 which depend on them.

The problem is then to determine the complex potentials $\phi(z_1)$, $\psi(z_2)$ for given boundary conditions. Green and Zerna [1] proceed, by way of the Hilbert problem, to concise representations of the solutions for the half plane (first, second, and third fundamental problems) and for the circular and elliptical hole in the infinite plane (first and second problems). They also give a solution of the hitherto untreated problem of the double hyperbolic notch in a plate in tension, obtained from complex potentials devised without Muskhelishvili methods.

Many of the papers which have recently appeared elsewhere have dealt with problems already disposed of in the earlier Russian papers. The next several paragraphs, based on Fridman's survey and the books of Lekhnitzki and Savin already mentioned, will indicate the general status of the subject. Aspects covered in Green and Zerna's book are omitted.

Elementary polynomials suffice for the two-dimensional problems of the uniform strip or slab in pure bending, "cantilever" bending, or uniform loading on one edge. Stress functions of the polar form $r^n \phi_n(\theta)$ serve for a wedge-shaped region (including the half plane) with vertex loading (force or couple), and for uniform and other loading on one edge. The simpler solutions, mostly two-dimensional, include: the pure bending of the partial circular ring, the bending by a force,* the tube with internal and external pressure—all with cylindrical orthotropy; the rotating elliptical disk, allowing variation of stress through the thickness; the orthotropic strip and half plane by Fourier methods. Lekhnitzki's [2] later book obtains a general form of solution for the half plane with given loading by Muskhelishvili's earlier method.

Savin employs a special method (1939) for the elliptical hole in the infinite plate, solved otherwise by Lekhnitzki in 1936. The detailed results given by Savin [1] are for the elliptical hole in tension at any angle, with general anisotropy (in two dimensions); the elliptical hole subject to uniform tangential force; uniform pressure over any part of the ellipse, with the concentrated force at any point as a limiting case; the elliptical hole at any orientation in a field of pure bending stress; and a "cantilever" stress field.

The second fundamental problem is treated similarly, with detailed results for a rigid elliptical inclusion subject to a couple in its plane, and as a disturbance in a uniform tension field at any angle. This last problem is also solved by Owens and Smith [1].

---

* Kosmodamianski [1] obtains the force solution for a less restricted anisotropy; the only plane of elastic symmetry of an element is parallel to the x-y plane.

Lekhnitzki [2] gives a perturbational solution for the plane problem for slightly anisotropic material, and evaluates the leading terms of the sequence of complex potentials for ovaloid holes disturbing uniform stress fields. A later paper (Lekhnitzki [3]) extends this analysis to plates with elliptic inclusions. Perturbational methods are further examined by Sokolnikoff [1].

Series solutions for the region inside an ellipse were given by Lekhnitzki (1937) and by Kufarev [1] and are included in the former's book [1] *Anisotropic Plates*. Kufarev [2] has expressed in double integral form the complex potentials for the infinite wedge region when both normal and tangential stresses are given on each straight edge. The uniformly stressed plane disturbed by a finite number of cuts on the x-axis was worked out by Mikhlin [1], and later by Green and Zerna [1] in terms of the Hilbert problem.

Savin [2, 3, 4] reduced the mixed boundary value problem of the rigid die indenting the orthotropic half plane (an elastic axis parallel to the edge) to the integral equation

$$(38) \qquad \int_{-l}^{l} P(t) \log |t - t_0| \, dt = f(t_0) + C$$

for the pressure $P(t_0)$ under the die $(-l < t_0 < l)$ which has the shape $f(t_0)$. This contains no reference to the anisotropy. It is the same as the equation governing the isotropic case, and the general solution is known (see, for instance, Muskhelishvili [1], Ch. 19). Savin's papers include tangential loading also. The corresponding problem for several dies (superposition is not valid) with Coulomb friction was solved by Galin [2] who reduced it to a Riemann-Hilbert problem. Isotropic problems of this kind are solved in Muskhelishvili's [1] book.

Lekhnitzki's [2] book contains his earlier theory of a generalized plane problem with displacements u, v, w independent of the co-ordinate z in an infinite cylinder with axis Oz, and arbitrary anisotropy. It requires two stress functions $F(x, y)$, $\Psi(x, y)$, satisfying a pair of linear differential equations of the fourth and third orders which produce three complex variables $z_i (i = 1, 2, 3)$ as arguments of six analytic functions $F_i(z_i)$, $\Psi_i(z_i)$. The applications are to an elliptic cylindrical cavity with forces on the cavity surface, free from axial components or axial variation, and to a parabolic cylinder with such forces.

Twisting and bending of bars, with their coupling through the anisotropy, are thoroughly treated. There is detailed attention to the elliptic and rectangular sections, and also, by an approximate variational method of Leibenson, to airfoil sections. Torsion of circular shafts of variable diameter (with cylindrical anisotropy) is covered by an extension of the Michell isotropic theory, and results found for the cone. Lekhnitzki gives the Saint-Venant flexure problem a general anisotropic formulation by a semi-inverse method, in terms of three complex functions $\Phi_i(z_i)$ holomorphic in the regions of the three z_i planes corresponding to the cross section.

For the symmetrical deformation of shapes of revolution with transverse isotropy (all directions in planes normal to the axis are elastically equivalent) the tangential displacement u_θ vanishes, and u_r, u_z are functions of r and z only. The shear stresses $\tau_{r\theta}$, $\tau_{\theta z}$ vanish. The other four stress components are expressed by Lekhnitzki [4] in terms of a single stress function satisfying a differential equation of the fourth order (a generalization of Love's isotropic function). Application is made to the semi-infinite solid with surface pressure, to the cylinder and to the heavy semi-infinite solid bounded by a horizontal plane, with a cylindrical bore on a vertical axis. New solutions in this class of "torsionless axisymmetry" have recently been obtained by Elliott [1], Shield [1], and Payne [1]. Eubanks and Sternberg [1] have established the completeness of Lekhnitzki's stress function. The two last papers provide references to round out this group, and give retrospective introductions. Their general accessibility renders further comment superfluous here, except to point out that Shield includes the nonsymmetrical problems of the flat elliptical crack in the infinite medium, the flat elliptical punch on the semi-infinite solid, and a distribution of face-parallel force on a thickness line in a slab. The distinctive elastic axis of the transversely isotropic material is normal to the elliptical area, or to the slab faces.

In general, it is scarcely to be expected that isotropic states can be brought into mathematical equivalence with anisotropic states, since the former are merely limiting forms of the latter. Such an equivalence has, however, been recognized in certain special problems: in connection with (38), and in Shield's [1] reduction of the general crack and punch problems (with transverse isotropy) to equivalent isotropic problems. The change from isotropic to anisotropic stress-strain relations in the linear theory is a form of linear transformation related to linear geometrical transformation [as z to z_1 and z_2 in (33) and (34)]. Lodge [1] has recently examined general linear transformations in the form

$$x_i \to x_i', \quad u_i \to u_i', \quad e_{ij} \to e_{ij}', \quad \sigma_{ij} \to \sigma_{ij}', \quad X_i \to X_i' \quad \text{(body force)}$$

where

$$x_i = a_{ri}x_r', \quad u_i' = a_{ir}u_r, \quad e_{ij}' = a_{ir}a_{js}e_{rs}, \quad \sigma_{ij} = a_{ri}a_{sj}\sigma_{rs}'$$

$$X_i = a_{ri}X_r'$$

The strain energy function

$$W = \tfrac{1}{2}c_{11}e_{11}{}^2 + c_{12}e_{11}e_{22} + \cdots$$

becomes

$$W = (\tfrac{1}{2}\lambda\delta_{ij}\delta_{rs} + \mu\delta_{ir}\delta_{js})e_{ij}{}'e_{rs}{}' \quad \text{(Kronecker } \delta\text{'s)}$$

appropriate to an isotropic solid of Lamé constants λ, μ. He finds that the taking over of isotropic solutions implies 14 conditions on the 21 constants of general anisotropy. For orthorhombic symmetry all but five are satisfied, and four independent constants remain. A Hertz contact result is transformed to apply to the four constant orthorhombic material. Conway [1] has given solutions for some of the previously solved elliptical hole problems by a linear geometrical transformation of circular hole results, both media being orthotropic. This suggests modification of Lodge's transformation, relaxing the requirement that one medium should be isotropic.

8. Thermal stress. Elastic waves induced by thermal shock

Accounts of most of the principal solved problems of thermal stress in the isotropic homogeneous solid are conveniently found in the book *Wärmespannungen* by Melan and Parkus (1953), in Ch. 14 of *Theory of Elasticity* by Timoshenko and Goodier (1951), and (as of 1937) in the older Russian monograph "Temperature Stress in the Theory of Elasticity" by Lebedev. The content of this monograph is covered by the two more recent books, with a few exceptions which include the nonsymmetrical case of the sphere, a short section on nonuniform elastic moduli, and the modification of the Kolosov formulas (1), (2), and (3) appropriate to nonuniform temperature (steady or unsteady). These become for temperature $T(z, \bar{z})$

$$(39) \qquad \sigma_x + \sigma_y = 2[\phi'(z) + \overline{\phi'(z)}] - T_1$$

$$(40) \qquad \sigma_y - \sigma_x + 2i\tau_{xy} = 2[\bar{z}\phi''(z) + \psi'(z)] - \int \frac{dT_1}{dz}\,d\bar{z}$$

$$(41) \qquad 2\mu(u + iv) = \kappa\phi(z) - z\overline{\phi'(z)} - \overline{\psi(z)} + \tfrac{1}{2}\int T_1\,dz$$

where $T_1 = E\alpha T/(1 - \nu)$, and α is the coefficient of expansion.

It is surprising to find the subject almost unrepresented in the subsequent Russian publications. Equations (39), (40), and (41) have been restated by Bogdanoff [1]. But the thermal stress problem is not essentially distinct from the ordinary surface and body-force problems of the unheated elastic solid. The reduction can be effected in several ways.* One of these employs a displacement potential which is the Newtonian potential of the temperature distribution, and in two dimensions the logarithmic potential. When in the two-dimensional case the temperature distribution is that of steady heat conduction in the homogeneous isotropic medium, the plane stress is that of a dislocation, or else zero. The dislocational examples of the circular region with an eccentric circular hole, or a plane with two holes, were worked out in detail in bipolar co-ordinates by Udoguchi [1, 2]. A problem solved otherwise by Gatewood [1] and by Hieke [1]—the circular cylinder with an eccentric circular cylindrical uniformly hot core—is readily reducible to a surface load problem of the solid cylinder by means of the logarithmic potential. This is the same as the eccentric shrink-fit problem with adhesion † during shrinkage, and appears in another guise as inserted oversize disks in several Russian papers employing the methods of Muskhelishvili and Sherman. Tarabasov [1, 2] takes first the eccentric (circular) shrink fit with adhesion for a single inserted disk of uniform oversize, then for several. But, the material being the same throughout, the latter reduces to the former by superposition. Ugodchikov [1] has two disks and an outer boundary of special mathematical form. In terms of the logarithmic potential all such problems can be regarded as solved wherever the first fundamental problem for the solid region enclosed by the outer boundary can be regarded as solved.‡

The disk (or cylinder with no axial variation) uniformly hot on one side of a chord, uniformly cold on the other (analyzed otherwise by Hieke [1]) is reduced by applying simple boundary compression parallel to the chord to the hot region to neutralize the free chordwise thermal expansion, then evaluating its removal from a complete unheated disk. The disk with a hot sector (also analyzed otherwise by Hieke [2]) can be reduced to a boundary force problem of the disk by way of the logarithmic potential of a sector.

There is also a general reduction to normal surface force proportional to T at the point, and body force derivable from T as a potential.§ It follows that certain general theorems, e.g., the Betti-Maxwell reciprocal theorem, will have thermoelastic counterparts which can be written down immediately. An earlier Russian paper, by Maisel [1], obtains a reciprocal theorem without regard to this possibility. Hieke [1],

---

* See, for instance, Timoshenko and Goodier, _op. cit._, Ch. 14.

† Contact with normal pressure only, a harder problem, is covered by Muskhelishvili [1].

‡ An application of the Muskhelishvili solution of the third (mixed boundary value) problem for the half plane is given by Huth [1].

§ Timoshenko and Goodier, _op. cit._, p. 422.

also treating the thermoelastic problem as distinct, finds as a general theorem that the total change of volume due to thermal stress (not including thermal expansion) is zero.

Temperature distributions which travel through the material as a wave appear in some recent papers. Melan [1] has determined the temperature distribution and the thermal stress for a disk which rotates, but has a spatially fixed temperature field, as in steady but spatially nonuniform hot gas flow past the disk. Moving heat sources in a fixed plate are considered by Melan [1, 2], and by other authors cited in a paper of Freudenthal and Weiner [1] on fatigue in terms of thermal stress produced by deformation.

When a sudden temperature rise is imparted to the surface of an elastic body, the thin heated skin is completely restrained from expanding in the surface by its attachment to the underlying mass. The initial state of stress is nonzero in the heated skin only, and immediately calculable. However, if a *finite amount* of heat is suddenly communicated to part of the surface, e.g., a circular area on the plane boundary of a semi-infinite solid, the thin heated disk under the restraint of the cold mass in which it is embedded exerts finite outward radial force in the surface. Sadowsky [1] has determined the stress so induced in this particular case, treating the problem as one of statics.

Other recent authors take, by implication, the view that sudden heating imparts particle velocities suddenly and therefore gives rise to an elastic wave with a sharp front. The problem then is essentially dynamical. Danilovskaya [1] and Mura [1] give similar independent investigations of this possibility, for the one-dimensional problem of the sudden (finite) temperature rise uniform over the whole plane surface of the semi-infinite solid. The ensuing temperature is given by a well-known error-function solution, and the equation for the elastic displacement incorporating this is integrated by Laplace transformation. Mura gives a particularly full account of the result, showing, in agreement with Danilovskaya, a finite stress discontinuity propagated from the initial thermal shock, and outrunning appreciable temperatures. This would imply that Sadowsky's problem may be essentially dynamical, the more so in that it requires an initially infinite temperature.

The sharp wave front has its origin in the initial temperature discontinuity at the surface. Danilovskaya [2] modifies the original problem by considering heating of the surface by sudden contact with a heat reservoir maintained at a constant temperature T_0, and providing a rate of heat flow proportional to the temperature difference $T - T_0$ at the surface. An elastic wave is propagated, but the sharp wave front is no longer present.

It would seem appropriate to examine further what order of magnitude the heat-transfer coefficient must have if the dynamical terms are to be significant at all. But this significance also involves the dimensions of the body. If the body is infinite in the direction of propagation, even gradual loading presents a dynamical problem—as in the semi-infinite rod under increasing end force—because the wave going in never turns back. If the body is bounded—as a short rod—the rapid reflections which occur bring about a quasi-statical response very quickly, before the load has had time to become appreciable. Thus, in a body of small size the rate of heat transfer requisite for significant dynamical effects will be greater than for a large body.

In problems of thermal stress, variation of coefficient of expansion α with temperature T raises no difficulty, the governing quantity being αT. Variation of elastic constants with T is taken into account in some recent papers. An effect which has not been taken into account, having been only very recently demonstrated as significant by Rosenfield and Averbach [1], is the change of α with stress: about 10% for a tensile stress of 40,000 psi in steel.

9. Three-dimensional contact problems

The state of the semi-infinite (isotropic, homogeneous) linearly elastic solid $(z > 0)$ under surface pressure only, $\sigma_z = -p(x, y)$, on some finite region Ω of the plane boundary, is representable by a single harmonic function $\phi_1(x, y, z)$. The stress σ_z is given by

$$(42) \qquad \frac{1 + \nu}{E} \sigma_z = \frac{\partial \phi_1}{\partial z} - z \frac{\partial^2 \phi_1}{\partial z^2}$$

and accordingly ϕ_1 is governed by the conditions

$$(43) \qquad \left(\frac{\partial \phi_1}{\partial z} \right)_{z=0} = -\frac{1 + \nu}{E} p(x, y) \quad \text{in } \Omega$$

$$(44) \qquad \qquad \qquad = 0 \qquad \qquad \text{outside } \Omega$$

This identifies ϕ_1 with the Newtonian potential of a simple surface layer on Ω, of surface density proportional to $p(x, y)$, so

$$(45) \qquad \phi_1(x, y, z) = -\frac{1 + \nu}{2\pi E} \iint_\Omega \frac{p(\xi, \eta) \, d\xi \, d\eta}{[(x - \xi)^2 + (y - \eta)^2 + z^2]^{\frac{1}{2}}}$$

The z-component of displacement at the surface is

$$(46) \qquad \qquad w = 2(1 - \nu)\phi_1$$

The Hertzian theory of contact of smooth elastic bodies rests on this basis.* The surfaces have originally nonzero curvature at the contact point O, and so each can be represented in its neighborhood in the form $z = ax^2 + by^2$ in terms of x, y in the tangent plane with an origin at O. The theory has found extensive application both in its statical form and in the pseudo-statical theory of impact which Hertz developed from it. Its application to predict the behavior of an aggregate of packed spheres would leave out the significant and interesting effects of friction. In the recent papers of Mindlin and his co-workers, friction is accounted for in solutions for two spheres with transverse forces and twisting couples. A comprehensive report of this development with its bibliography, and an encouraging connection with experimental results, is generally available (Mindlin [1]). Slip occurs on an outside annulus of the circular contact; adhesion with friction prevails within.

In the Russian publications the problem of the *rigid* stamp or die, enforcing without friction a prescribed vertical displacement and leaving free the rest of the plane surface of the semi-infinite solid, has occupied a prominent place since 1935. It forms the main content of the second half † of the book by Galin [1]. As may be seen from (46), the potential ϕ_1 is prescribed within the contact since w is, and outside the contact it must have $(\partial\phi_1/\partial z)_{z=0} = 0$ by (44). The actual pressure $p(x, y)$ under the stamp is to be found, essentially from (45), as an integral equation. If Ω is regarded as a lamina in otherwise empty space, the potential ϕ_1 is required to approach a prescribed value, say $f(x, y)$, on this lamina from above or below (the given stamp form), to be continuous throughout the space, and to vanish at infinity—the Dirichlet problem for the space outside a thin lamina of shape Ω. The condition (44) will be satisfied by symmetry. The solutions given by Galin, in the main drawn from his own papers, extend to circular, elliptic, and (infinite) wedge shapes for Ω. The form $f(x, y)$ of the prescribed displacement w within Ω—the form of the base surface of the rigid stamp—is arbitrary in the first instance.

When Ω is circular, a result of Hobson [1] provides the required potential in spheroidal co-ordinates. The force P required to maintain the displacement $f(x, y)$, or in polar co-ordinates ρ, θ on $z = 0$, $f(\rho, \theta)$, follows almost immediately from the limiting form of the potential at infinity as

$$(47) \qquad P = \frac{E}{\pi(1 - \nu^2)} \int_0^{2\pi} \int_0^a \frac{\rho f(\rho, \theta)}{(a^2 - \rho^2)^{1/2}} \, d\rho \, d\theta$$

* See, for instance, A. E. H. Love, *The Mathematical Theory of Elasticity*, Cambridge University Press, Cambridge, England, Ch. 8, 1927.

† The first half, on the two-dimensional problems, was outlined in Section 5 above.

where a is the radius of the circle. This is a generalization, obtained by Galin [1] in 1946, of the special case $f(\rho, \theta) =$ constant given by Boussinesq (1885) in *Application des potentiels* .... . The pressure distribution under the stamp is approached through a Green's function for a point charge within the flat empty region in a hole of shape Ω in an infinite conducting plate at zero potential. The function used was obtained by Kochin * (1941) from an earlier, less simple, result of Sommerfeld. The pressure for the general $f(\rho, \theta)$ is not worked out. For the axisymmetric case it is given as

$$(48) \qquad p(\rho) = -\frac{E}{4\pi(1 - \nu^2)} \int_0^a \Delta f(\rho_1) H(\rho, \rho_1)\, d\rho_1$$

where Δ is the Laplacian operator in ρ_1, $\partial^2/\partial\rho_1^2 + \rho_1^{-1}\, \partial/\partial\rho_1$. The kernel H is given as

$$H(\rho, \rho_1) = \frac{2}{\pi} \int_0^{2\pi} \rho_1(\rho^2 - 2\rho_1 \cos\theta_1 + \rho_1^2)^{-\frac{1}{2}}$$

$$\times \text{ arc tan } [a^{-1}(a^2 - \rho^2)^{\frac{1}{2}}(a^2 - \rho_1^2)^{\frac{1}{2}}(\rho^2 - 2\rho\rho_1 \cos\theta_1 + \rho_1^2)^{-\frac{1}{2}}]\, d\theta_1$$

The displacement c of the stamp (from zero load contact) is then found as

$$c = -\int_0^a \Delta f(\rho_1) \rho_1 \text{ arc tanh } \left(1 - \frac{\rho_1^2}{a^2}\right)^{\frac{1}{2}} d\rho_1$$

This axisymmetric case is also completely worked out by Leonov * (1939), and in Sneddon's *Fourier Transforms* (*op. cit.*) (from Harding and Sneddon [1]), by means of Hankel transforms. Another form of solution is given by Green and Zerna [1] (p. 172).

The special case $f(\rho) = A\rho^\lambda$, previously solved by Shtaerman and Lourie, is evaluated in gamma functions of λ, and the force-displacement ratio exhibited in Galin [1] (Fig. 40) as a curve for $1 < \lambda < 4$.

The results cited above are in the first instance for a smooth prescribed displacement $f(x, y)$ or $f(\rho)$, and the radius a of the contact circle may have to be found from them if P is given rather than a. If $f(\rho)$ is taken as a constant c they merge into the well-known Boussinesq result for the flat-based circular stamp. If the nonflat circular stamp is pressed with just sufficient force to effect contact over the whole base area, further force will simply superpose this Boussinesq state to the extent measured by the further displacement.

The surface $z = 0$ of the semi-infinite solid has been supposed free, except for the pressure under the stamp. If there are outside forces this pressure is modified. For the flat circular stamp (force P) in the

* See Galin [1].

presence of a concentrated normal force Q on $z = 0$ at $x = l$, $y = 0$, Galin has

$$p(x, y) = -\frac{P}{2\pi(a^2 - x^2 - y^2)^{1/2}} + \frac{Q}{\pi^2[(x - l)^2 + y^2]} \cdot \left(\frac{l^2 - a^2}{a^2 - x^2 - y^2}\right)^{1/2}$$

and since the flat base of the stamp is required to remain level, the line of action of P follows and will not be central.

The effects of friction in axisymmetric contact are included (Galin [1]) only as to the steady revolving stamp. Limiting frictional forces in the circumferential direction prevail over all the circular contact area, but the coefficient of friction is taken as a function $F(\omega\rho)$ of the rubbing velocity $\omega\rho$.

The treatment of this problem is facilitated by the fact that when there is no dependence on θ (in cylindrical co-ordinates ρ, θ, z) the equations of linear elasticity in terms of displacements (Lamé equations) subdivide into two independent sets.* The first set (two equations) involve u_ρ and u_z (formerly w), and if u_θ is taken zero, leading to $\tau_{\rho\theta} = \tau_{z\theta} = 0$, they suffice for stress distributions involving nonzero σ_ρ, σ_θ, σ_z and $\tau_{\rho z}$—torsionless axisymmetry. The frictionless stamp problem is of this type. The second set (one equation only) involves only u_θ, and if u_ρ, u_z are taken zero, leading to $\sigma_\rho = \sigma_\theta = \sigma_z = \tau_{\rho z} = 0$, it suffices for problems of the semi-infinite solid with ringwise shear $\tau_{z\theta}$ on the surface $z = 0$ (as well as the Michell torsion theory for solids of revolution). The friction forces under the revolving stamp present a problem of this kind; other examples are the torsion of pressed spheres with friction (Lubkin [1]), and, when dynamical terms are included, the Reissner-Sagoci [1] solution for the torsional oscillations of the semi-infinite solid forced by a rigid disk. Consequently, the pressure $p(\rho)$ under the stamp is not altered by the friction. It is given by the frictionless result (48), and determines the distribution of frictional force $\tau_{z\theta} = F(\omega\rho)p(\rho)$ and the required turning moment M. For the coefficient of friction $F(\omega\rho)$ a constant, μ, Galin [1] gives for the revolving circular stamp

$M = \frac{1}{4} \cdot \pi\mu aP$ for a flat base of radius a

$M = \frac{3}{16} \cdot \pi\mu aP$ for a base in the form of a paraboloid of revolution (Hertz case), contact

radius $a = \left(\frac{3}{4}\frac{1 - v^2}{E} PR\right)^{1/3}$, R

= radius of curvature of generating parabola at vertex

For uniform pressure distribution the value is $M = \frac{2}{3} \cdot \mu aP$.

* E. Reissner, *Ing.-Arch.*, **8**, 229–245, 1937.

Returning to the frictionless stamp, an *elliptic plan form* requires solution of the Dirichlet problem for the elliptic disk, the potential taking the assigned value $f(x, y)$ on it from above and below. The The earliest results appear to be those of Lourie [4, 5, 6] (1939, 1940, 1941). Galin's [1] solution (1947) employs ellipsoidal co-ordinates ν, μ, ρ and the functions of Lamé which, in triple products, provide harmonic functions in space. It relies on the chapter "Ellipsoidal harmonics and Lamé's equation" in Whittaker and Watson's *Modern Analysis*. The properties of these functions furnish the result that when the equation of the base surface $z = f(x, y)$ has the form $z = P_n(x, y)$, P_n being a polynomial of degree n, the pressure under the stamp will have the form

$$p(x, y) = \left(1 - \frac{x^2}{a^2} - \frac{y^2}{b^2}\right)^{-\frac{1}{2}} P_n^*(x, y)$$

where $P_n^*(x, y)$ is another polynomial of degree n. Galin shows that the central force P and the moments M_x, M_y about Ox, Oy required for the displacement $f(x, y)$ under a stamp of any plan form can be found from the asymptotic forms of the potential $\phi(x, y, z)$ for remote points, and of its derivatives $\partial\phi/\partial y$, $\partial\phi/\partial x$ in the planes $y = 0$, $x = 0$ respectively. With ϕ taken in terms of the Lamé functions $E_n{}^m$, $F_n{}^m$ (Whittaker and Watson, Ch. 23) as

$$\phi(x, y, z) = \phi^*(\nu, \mu, \rho) = \sum_{n=0}^{\infty} \sum_{m=0}^{2n+1} A_{nm} E_n{}^m(\nu) E_n{}^m(\mu) F_n{}^m(\rho)$$

the evaluation of the asymptotic forms required is the evaluation of the coefficients A_0, A_1, A_2 in

$$\phi = A_0 \cdot \frac{1}{s} + \frac{1}{3} A_1 \frac{x}{s^3} + \frac{1}{3} A_2 \frac{y}{s^3} + \cdots$$

where s is $(x^2 + y^2 + z^2)^{\frac{1}{2}}$ and is large. The orthogonality properties of the Lamé functions lead to expressions for these coefficients, and so to the following results. For the force on the elliptic stamp

$$(49) \quad P = \frac{E}{2(1 - \sigma^2)\Psi_0(1)} \cdot \frac{1}{b} \iint_\epsilon f(x, y) \left(1 - \frac{x^2}{a^2} - \frac{y^2}{b^2}\right)^{-\frac{1}{2}} dx\, dy$$

Here σ is Poisson's ratio, E Young's modulus, ϵ the interior of the ellipse $x^2/a^2 + y^2/b^2 = 1$, and $\Psi_0(1)$ is $F(\pi/2, e)$, the complete elliptic integral of the first kind [$b^2 = a^2(1 - e^2)$]. Equation (47) for the circular stamp is a special case of (49). For the moments required by the elliptic stamp *

$$M_y = \frac{E}{2(1-\sigma^2)} \cdot \frac{1}{b} \cdot \frac{1}{\Psi_1^{(1)}(1)} \iint_\epsilon xf(x,y) \left(1 - \frac{x^2}{a^2} - \frac{y^2}{b^2}\right)^{-\frac{1}{2}} dx\,dy$$

$$M_x = \frac{E}{2(1-\sigma^2)} \cdot \frac{1}{b} \cdot \frac{1}{\Psi_1^{(2)}(1)(1-e^2)}$$

$$\times \iint_\epsilon yf(x,y) \left(1 - \frac{x^2}{a^2} - \frac{y^2}{b^2}\right)^{-\frac{1}{2}} dx\,dy$$

where

$$\Psi_1^{(1)}(1) = \frac{1}{e^2}\left[F\left(\frac{\pi}{2}, e\right) - E\left(\frac{\pi}{2}, e\right)\right]$$

$$\Psi_1^{(2)}(1) = \frac{1}{e^2}\left[\Pi\left(\frac{\pi}{2}, -e^2, e\right) - F\left(\frac{\pi}{2}, e\right)\right]$$

and F, E, Π are complete elliptic integrals of the first, second, and third kinds.

Steps towards a solution for the *frictionless stamp of arbitrary plan form* have been taken by Mossakowskii [2] and Galin [1]. The latter gives upper and lower bounds for the force P required for a given displacement of a stamp with a flat base, which remains parallel to the undeformed surface. For the upper bound the actual contact region Ω on $z = 0$ is replaced by the circumscribing ellipse, of such orientation as to have smallest area. Then formula (49) is used. The lower bound is provided by one of the symmetrization inequalities established by Pólya and Szegö.† The electrostatic equivalent of the frictionless stamp problem [cf. (45)] makes the capacity of the thin disk Ω analogous to the force-displacement ratio (P/δ) of the stamp, provided the base is flat and remains level. The symmetrization inequality makes the capacity of the flat circular disk (as a limiting form of a solid ellipsoid of revolution) less than that of any other lamina of the same area. Accordingly Galin obtains the lower bound for the stamp force by taking a flat-based round stamp of base area equal to that of the given stamp (Ω). Applied to the square plan form (side $2h$), the circumscribing ellipse

---

* The denominator factor $(1 - e^2)$ in M_x is missing in Galin's expression, as Mossakowskii [1] pointed out.

† *Amer. J. Math.*, **67**, 1–32, 1945.

becomes the circle of radius $\sqrt{2}h$, and the circle of equal area has radius $(2/\sqrt{\pi})h$. Then

$$2.26\,\frac{Eh\delta}{1-\sigma^2} < P < 2.82\,\frac{Eh\delta}{1-\sigma^2}$$

A solution for the frictionless stamp of *sectorial plan form*, occupying the region $-\alpha < \theta < \alpha$ of $z = 0$ and infinite in the radial direction, was constructed by Galin (1947). It rests on an observation of Hobson [2] that a harmonic function Φ which in spherical co-ordinates r, θ, ϕ is independent of r will satisfy

$$\frac{\partial^2 \Phi}{\partial \xi^2} + \frac{\partial^2 \Phi}{\partial \eta^2} = 0 \qquad \text{where } \xi = \tan\frac{\theta}{2}\cos\phi$$

$$\eta = \tan\frac{\theta}{2}\sin\phi$$

The analysis proceeds in the complex plane of $\zeta = \xi + i\eta$. The lines of constant pressure are drawn for sectors including angles $\pi/2$ and $\pi/4$, for the region of the vertex (where a singularity occurs). The solution is given in Galin's [1] book, which also contains some considerations on the beam on the elastic semi-infinite solid, and on a rigid elliptical paraboloid (Hertz local shape) pressed centrally on a thin circular plate. The book has a distinctive historical introduction, a Russian bibliography of 109 items, and a non-Russian bibliography of 17 items. For the problems of the beam and the plate resting on the elastic semi-infinite solid 20 Russian papers are referred to, and a book by Gorbunov-Posadov, *Beams and Plates on the Elastic Half-Space* (1949).

The effects of friction or adhesion in the three-dimensional problems have received little attention in the Russian publications. The revolving stamp has been referred to above. Mossakowskii [2, 3] considers the fundamental mixed boundary value problem, invokes the Hilbert problem (Section 5 above), and treats the axisymmetric stamp with adhesion as an example.

Recent solutions of stamp problems (circular and elliptic plan form) for certain anisotropic semi-infinite solids appear in papers of Shield [1], Payne [1], and their references.

10. Wave propagation. Traveling loads and sources of disturbance

If a load of unchanging form (e.g., uniform pressure on a surface area of unchanging size and shape) travels at uniform velocity V along the plane surface of a semi-infinite solid, it may be imagined to carry along with it an unvarying pattern of deformation—a wave. This steady régime is a *forced wave*, analogous to steady forced vibration. The transient régime, say from starting up the load, is left out of account. The implied justification is slight damping, as in vibration theory. A number of recent investigations in several countries deal with this forced wave problem.

The infinite (isotropic, homogeneous) medium has its two velocities of propagation, $a = [(\lambda + 2\mu)/\rho]^{\frac{1}{2}}$ for dilatational (irrotational) waves, and $b = (\mu/\rho)^{\frac{1}{2}}$ for transverse (shear, equivoluminal) waves. Forced wave motions resulting from sources of disturbance traveling at uniform velocity V may be conveniently classified as

$$\text{(50)} \qquad \begin{aligned} &subsonic &&\text{when} \quad 0 < V < b \\ &intersonic &&\text{when} \quad b < V < a \\ &supersonic &&\text{when} \quad a < V < \infty \end{aligned}$$

For plane stress a would be $[E/\rho(1 - \nu^2)]^{\frac{1}{2}}$.

The source of disturbance may be held fixed, in fixed axes x', y', z', and the elastic medium allowed to flow past it. We can have $x' = x + Vt$, $y' = y$, $z' = z$, where x, y, z refer to axes traveling with the medium. Displacement u', v', w' can be defined, like velocity in Eulerian hydrodynamics, as that observed at a fixed point x', y', z' as the medium moves past: it is the displacement of the point which would have been at x', y', z' in the absence of deformation. The equations of motion on this basis are (without restriction to steady motion of the medium)

$$(\lambda + \mu) \left(\frac{\partial}{\partial x'}, \frac{\partial}{\partial y'}, \frac{\partial}{\partial z'} \right) \Delta + \mu \nabla^2 (u', v', w') = \rho \left(V \frac{\partial}{\partial x'} + \frac{\partial}{\partial t} \right)^2 (u', v', w')$$

where $\Delta = \partial u'/\partial x' + \partial v'/\partial y' + \partial w'/\partial z'$ and ∇^2 is the Laplacian operator in x', y', z'. The displacement can be taken in general as grad ϕ + curl (ψ_1, ψ_2, ψ_3). The scalar potential ϕ of the irrotational part, and the vector potential (ψ_1, ψ_2, ψ_3) of the equivoluminal part, satisfy the equations

$$\nabla^2 \phi = \frac{1}{a^2} \left(V \frac{\partial}{\partial x'} + \frac{\partial}{\partial t} \right)^2 \phi; \qquad \nabla^2 \psi_i = \frac{1}{b^2} \left(V \frac{\partial}{\partial x'} + \frac{\partial}{\partial t} \right)^2 \psi_i$$

For a steady, forced wave the time dependence disappears and the equations are

$$\text{(51)} \qquad \left(1 - \frac{V^2}{a^2} \right) \frac{\partial^2 \phi}{\partial x'^2} + \frac{\partial^2 \phi}{\partial y'^2} + \frac{\partial^2 \phi}{\partial z'^2} = 0$$

$$\left(1 - \frac{V^2}{b^2} \right) \frac{\partial^2 \psi_i}{\partial x'^2} + \frac{\partial^2 \psi_i}{\partial y'^2} + \frac{\partial^2 \psi_i}{\partial z'^2} = 0$$

but these follow immediately from the usual equations of motion in x, y, z and the fact that ϕ and ψ_i are now functions of x', y', z' only. The spatial problems in x', y', z' which they present will have elliptic character in both parts when subsonic [by the classification (50)]. Intersonic motion has an elliptic ϕ-equation but hyperbolic ψ-equations. Supersonic motion is hyperbolic in both parts.

In a paper on the two-dimensional forced wave problem Sneddon [1] has shown that traveling normal or shear loading on the bounding line of the semi-infinite plane can be treated in an elementary way,* at least for the subsonic range. A twice differentiable function $f(x - Vt \pm iky)$ is a solution of $a^2 \nabla^2 f = \partial^2 f/\partial t^2$ provided $k^2 = 1 - (V^2/a^2)$. With

$$u = \frac{\partial \phi}{\partial x} + \frac{\partial \psi}{\partial y}, \qquad v = \frac{\partial \phi}{\partial y} - \frac{\partial \psi}{\partial x}$$

(and $w = 0$), ϕ and ψ may be given the real forms

$$\phi = \operatorname{Re} f(x - Vt + i\alpha y)$$

$$= f(x - Vt + i\alpha y) + \bar{f}(x - Vt - i\alpha y), \qquad \alpha^2 = 1 - \frac{V^2}{a^2}$$

$$\psi = \operatorname{Re} g(x - Vt + i\gamma y)$$

$$= g(x - Vt + i\gamma y) + \bar{g}(x - Vt - i\gamma y), \qquad \gamma^2 = 1 - \frac{V^2}{b^2}$$

where bars indicate conjugate forms. These represent a steady forced wave, α and γ being real. They are found to satisfy the boundary conditions

$$\tau_{xy} = 0, \qquad \sigma_y = -\tfrac{1}{2}[F''(x - Vt) + \bar{F}''(x - Vt)] \qquad \text{on } y = 0$$

if

$$f(z) = -\frac{1 - \tfrac{1}{2}\beta_2{}^2}{4\mu(\theta + \chi)} F(z), \qquad g(z) = \frac{2i\alpha}{1 + \gamma^2} f(z), \qquad z = x + iy$$

$$\theta = -\tfrac{1}{2}(1 - \tfrac{1}{2}\beta_2{}^2), \qquad \chi = (1 - \beta_1{}^2)^{1/2}(1 - \beta_2{}^2)^{1/2}$$

$$\beta_1 = \frac{V}{a}, \qquad \beta_2 = \frac{V}{b}$$

For a concentrated force P, Sneddon finds $F(z) = -\pi^{-1}iP(z \log z - z)$. The traveling shear load problem is solved in the same fashion.

---

* Fourier transform solutions are given in Sneddon's *Fourier Transforms*, p. 444, and, for further problems of moving (also periodic and impulsive) forces within the infinite medium in two and three dimensions by G. Eason, J. Fulton, and I. N. Sneddon, *Phil. Trans. Roy. Soc. (London)*, A, **248**, 575–607, 1956.

This solution is in terms of functions of the two complex variables $x - Vt + i\alpha y$, $x - Vt + i\gamma y$, and evidently problems of this kind will have a close relation to the statical problems of the anisotropic medium represented by (35), (36), and (37). It seems probable that there would be a corresponding simple solution for the forced wave in the anisotropic medium, in terms of two complex variables involving both the velocity V and the anisotropy.

Somewhat more complicated analysis on the same general basis, but involving nonelementary solution of functional equations, is employed by Zvolinski, for two-dimensional problems of the elastic semiinfinite solid supporting a fluid layer in which the motion is acoustic. The problems are steady propagation of free waves, parallel to the surface (Zvolinski [1]) and the motion following a concentrated impulsive disturbance in the solid (Zvolinski [2, 3]).

There would presumably be no difficulty in solving Sneddon's problem when traveling boundary displacements are prescribed instead of traveling loads. But the mixed problem is not likely to yield to elementary methods. It is treated by Galin (1943), and included in his book (Galin [1]). The mixed boundary conditions are satisfied for the (subsonic) motion of a moving rigid stamp of assignable form $y = f(x)$, with Coulomb friction, pressed into the edge of the half plane by a normal force P. This is done by way of the Riemann-Hilbert problem indicated in Section 5 in connection with the stationary stamp. As an example, Galin finds that the normal pressure distribution under the level flat stamp $(-l < x < l$ for $t = 0)$ is

$$p(x) = -\frac{P}{\pi} \sin \pi\theta \cdot \frac{1}{(l^2 - x^2)^{1/2}} \left(\frac{l + x}{l - x}\right)^{1/2 - \theta}$$

where

$$\theta = \frac{1}{\pi} \arctan \frac{-\dfrac{m^2}{2\mu}\left(1 - \dfrac{1 - 2\nu}{2 - 2\nu} m^2\right)}{\left[\left[1 + \left(\dfrac{3}{2} - 2\nu\right) m^2\right]\left(1 - \dfrac{1}{2} m^2\right)\left(1 - \dfrac{1 - 2\nu}{2 - 2\nu} m^2\right)^{1/2}\right.}{\left. -\left(1 - \dfrac{1 - 2\nu}{2 - 2\nu} m^2\right)\left[1 + (2 - 2\nu)m^2\right]\left(1 - m^2\right)^{1/2}\right]}$$

and $m = V/b$. Here μ is the coefficient of friction.

Supersonic flow of an elastic medium past a cone gives rise to two "Mach cones," one for each of the two fundamental velocities of propagation a and b. Solutions for a cone and a wedge are given by Kusukawa [1].

The theory of traveling sources of disturbance in the elastic medium is of current interest in the study of dislocations in crystals. Expositions and references are given by Mott [1] and Nabarro [1]. Eshelby [1] examines the notion of force on a singularity of an elastic stress field, implied by the possibility of a decrease of potential energy through a shift of the singularity. Saenz [1] investigates uniformly traveling dislocations in an anisotropic homogeneous medium, with generalization of the statical Weingarten-Volterra theorems.

11. Diffraction. Pulse propagation

A second group of papers on two-dimensional wave motion is concerned with *diffraction* of a plane wave train or pulse by a semi-infinite slit with free or fixed edges. This is the counterpart in elasticity theory of the well-known Sommerfeld diffraction problem—a semi-infinite screen in electromagnetic wave propagation. It seems that no successful adaptation of Sommerfeld's method to the elastic problem—with its two boundary conditions instead of one—has been achieved. Other methods have been employed in electromagnetic screen diffraction problems, and references to these with notes on their adaptability to the elastic problems may be found in a paper by Maue [1]. A method previously used for electromagnetic problems (superposition of plane waves in all directions) was also used by Sauter [1] to obtain a solution for the elastic half space with surface forces assignable both as to spatial distribution and as to time (Lamb's problem in two dimensions). Distinct methods are introduced by Maue [2] in a later paper, and by a group of Russian investigators. These will now be considered in some detail and the connection between them pointed out.

In Maue's problem an infinite plate (x, y) is under statical uniform tension $\sigma_y^{(1)}$. At $t = 0$ a semi-infinite straight transverse cut is made instantaneously on the negative x-axis, so for $t > 0$ the faces of the cut are free. The point of departure is a similarity consideration, which may be roughly translated as "Since the problem contains no characteristic length, the spatial distribution of stress is the same at all subsequent times, if the unit of length is proportional to t."

There is a class of diffraction problems governed by the wave equation in two dimensions, and devoid of characteristic length. It includes the semi-infinite slit, screen, barrier, at which traction, displacement, velocity, or other quantity vanishes, in the way of an incident pulse in the form of a step function $u_0 H(\xi)$, where $H(\xi) = 1$ for $\xi > 0$, 0 for $\xi < 0$. The direction of incidence is Ox' at any angle α to the slit, so $\xi = x' - at$, where a is the velocity of propagation. The subsequent states are given by $u(x, y, t)$ where $a^2 \nabla^2 u = \partial^2 u / \partial t^2$, and u is intended to be of the same nature as u_0 (if u_0 is a velocity, u is a velocity component). Then the solution for u would be a relation between the seven quantities,

$$u, u_0, r, \theta, t, a, \alpha$$

where r and θ are polar co-ordinates centered on the end of the slit. These seven quantities require two fundamental units for their specification—units of length and time, if u and u_0 can be so specified, as when they are velocities. There are consequently five independent dimensionless groups, and the solution must be conformable to

$$\frac{u}{u_0} = f_1\left(\frac{at}{r}, \theta, \frac{u_0}{a}, \alpha\right)$$

The problem is linear, so u must be proportional to u_0, and the argument u_0/a must drop out. Then

$$\frac{u}{u_0} = f_2\left(\frac{at}{r}, \theta, \alpha\right)$$

There must be solutions of $a^2 \nabla^2 u = \partial^2 u/\partial t^2$ which have this form. Writing z for at/r, the function $u(z, \theta)$ must be a solution of

$$(z^2 - 1)\frac{\partial^2 u}{\partial z^2} + z\frac{\partial u}{\partial z} + \frac{\partial^2 u}{\partial \theta^2} = 0$$

and the change of variable $z = \cos \chi$ reduces this to

$$\partial^2 u/\partial \chi^2 = \partial^2 u/\partial \theta^2$$

i.e., to the one-dimensional wave equation. Then

$$(52) \qquad u = f(\theta + \chi) + g(\theta - \chi), \qquad \cos \chi = \frac{at}{r}$$

The variable χ is real if $r > at$ and imaginary if $r < at$. This is Maue's form of solution. In the elastic problem there are two functions ϕ and ψ satisfying $a^2 \nabla^2 \phi = \partial^2 \phi/\partial t^2$, $b^2 \nabla^2 \psi = \partial^2 \psi/\partial t^2$. Each may then be taken in the form (52). They are coupled through the boundary conditions on the cut, which reduce the problem to functional equations. Satisfying these conditions Maue is led to integral equations in one complex plane to determine the part of the solution deriving from ϕ, and in a second complex plane for the part deriving from ψ.

The essence of this problem is of course the sudden annihilation of the original tensile stress along the cut. If this is taken in conjunction with an incident pulse of tension, it will give a solution for the diffraction of such a pulse when it encounters a slit. Evidently the pulse must travel in the y-direction, normal to the slit, so that instantaneous annihilation of its tensile stress is required all along the slit simultaneously. The general angle of incidence is covered by Fridman [3].

The group of Russian investigators reduce their diffraction and other wave problems to readily solvable Hilbert problems (see Section 5 above) in two auxiliary complex planes. The point of departure here is a solution of the two-dimensional wave equation $a^2 \nabla^2 \phi = \partial^2 \phi / \partial t^2$ in the form $\phi = \text{Re } \Phi(\Theta)$, where Θ is a function of x, y, t defined by

$$(53) \qquad t - \Theta x - \sqrt{a^{-2} - \Theta^2} \cdot y = 0$$

This is ascribed to Smirnov and Sobolev [1]. It turns out to be essentially the same as Maue's form (52). If in (52) we take the functions $f(\theta + \chi)$, $g(\theta - \chi)$ as $F[\cos(\theta \pm \chi)]$, the argument $\cos(\theta \pm \chi)$ can be identified with the Θ of (53). We have

$$(54) \quad \cos(\theta \pm \chi) = \cos\theta\cos\chi \mp \sin\theta\sin\chi = \frac{x}{r} \cdot \frac{at}{r} \mp \frac{y}{r}\sqrt{1 - \frac{a^2 t^2}{r^2}}$$

But squaring out the radical in (53), and solving the resulting quadratic for Θ yields exactly the last member of (54).

Within this group Fridman [2, 3] considers an incident pulse of step form traveling at velocity a (if a longitudinal wave) or b (if a transverse wave) in an arbitrary direction in the x-y plane. The displacement components for the longitudinal pulse are

$$u = -cH(t - cx + \sqrt{a^{-2} - c^2} \cdot y)$$

$$v = \sqrt{a^{-2} - c^2} \cdot H(t - cx + \sqrt{a^{-2} - c^2} \cdot y)$$

where c is a real direction number, and $0 < c < a^{-1}$. This pulse encounters the end ($x = y = 0$) of a semi-infinite slit $y = 0$, $x > 0$, at time $t = 0$ and is then diffracted by the slit. The displacements are taken as $u = u_1 + u_2$, $v = v_1 + v_2$ in dependence on two functions Θ_1, Θ_2 of the same kind as Θ in (53):

(55)

$$u_1(x, y, t) = \text{Re } U_1(\Theta_1) \qquad\qquad v_1(x, y, t) = \text{Re } V_1(\Theta_1)$$

$$u_2(x, y, t) = \text{Re } U_2(\Theta_2) \qquad\qquad v_2(x, y, t) = \text{Re } V_2(\Theta_2)$$

$$t - \Theta_1 x - \sqrt{a^{-2} - \Theta_1^2} \cdot y = 0 \qquad\qquad t - \Theta_2 x - \sqrt{b^{-2} - \Theta_2^2} \cdot y = 0$$

The equations $\partial u_1/\partial y - \partial v_1/\partial x = 0$ of the irrotational part, and $\partial u_2/\partial x + \partial v_2/\partial y = 0$ of the equivoluminal part imply

(56)
$$\left. \begin{aligned} \sqrt{a^{-2} - \Theta_1{}^2} \cdot U_1'(\Theta_1) + \Theta_1 V_1'(\Theta_1) &= iC_1 \\[2mm] \Theta_2 U_2'(\Theta_2) + \sqrt{b^{-2} - \Theta_2{}^2} \cdot V_2'(\Theta_2) &= iC_2 \end{aligned} \right\} \quad C_1, C_2 \text{ real constants}$$

The functions $U_1(\Theta_1)$, $V_1(\Theta_1)$ are holomorphic in the plane of $\Theta_1 = \xi_1 + i\eta_1$ cut along $\xi_1 > -a^{-1}$. For $U_2(\Theta_2)$, $V_2(\Theta_2)$ in the Θ_2 plane the cut is along $\xi_2 > -b^{-1}$. These four functions, linked by (56), are expressed in terms of two others. Application of the conditions at the slit, which imply discontinuities from one face to the other, produces two Hilbert problems, one for each function. The line of discontinuity is in each case the segment $-b^{-1} < \xi < -a^{-1}$ of the real axis in the Θ-plane [on the x-axis of the physical plane $\Theta_1 = \Theta_2 = t/x$ by (55)], and accordingly the results are expressed as Cauchy integrals on this segment. Two problems are solved: (a) in Fridman [2] the two faces of the slit are completely fixed, (b) in Fridman [3] the two faces of the slit are free. Photoelastic photographs of a case of this second problem —the pulse front is normal to the slit, the direction of propagation parallel—have been made by Christie [1], but not in connection with this theory.

The same method is applied by Sveklo [1] to a mixed boundary value problem of the semi-infinite plane. An instantaneous impulse of tangential load is imparted at a point ($x = y = 0$) of the edge $y = 0$. Otherwise the shear stress τ_{xy} on the edge is permanently zero everywhere. The positive half of the edge is also free of normal load σ_y, but the negative half is allowed no normal displacement ($v = 0$). The solution is obtained from a single Hilbert problem on the interval $-b^{-1} < \xi < -a^{-1}$ of the Θ-plane.

The two-dimensional problem of a sudden rise of pressure (or suddenly imparted radial velocity) in a circular hole in an infinite elastic medium * has been treated twice by Laplace transforms. Kromm [1] reduced it to an integral equation and obtained representative curves by numerical solution. It has since been pointed out that the corresponding rotary problem—uniform tangential force in the hole instead of pressure (or tangential instead of radial velocity)—is closely analogous (as in the statical case) and has a similar solution.† In an independent

* The acoustic problem for a fluid is simpler. The solution was given by H. Lamb; cf. *Hydrodynamics*, Cambridge University Press, Cambridge, England, Art. 302, p. 524, 1932.

† (*Added in proof*) J. N. Goodier and W. E. Jahsman, *J. Appl. Mech.*, **23**, 284–286, 1956.

analysis Selberg [1] has obtained results for the pressure problem more directly by an evaluation of the inversion integral. His paper also contains a solution for sudden pressure rise in a spherical cavity.*

In these pressure problems there is only one wave velocity involved since the motion is clearly irrotational, and hence a simple net of characteristics. Finite difference methods on such a net have readily yielded extensive information on the thick-walled cylinder with a sudden rise of internal pressure. The behavior of the wave as it moves outwards, and also after it has undergone reflection at the free outer surface, is exhibited in curves in a paper of Agababyan [1].

12. Seismic and vibrational problems

The basic paper of Lamb [1] (1904) on the dynamics of the semi-infinite solid has given rise to a group of papers in Russian, and an independent group in the western languages. Some spring from seismological questions, as did Lamb's, and others from technical problems such as impact recording or foundation disturbances from vibrating machines. In both respects experimental connections are developing.

A paper by Gutin [1] refers to earlier work by Sobolev [1, 2], Smirnov and Sobolev [1], Narishkina [1], and Sherman [4] dealing with evaluations for the interior of the solid. Petrashen [1, 2] has extended a recent reinvestigation of the two-dimensional problem to the elastic slab, with a concentrated impulsive surface disturbance, and is able to show in curves a characteristic difference due to the finite thickness: the surface motion acquires a damped oscillatory character. Gutin shows that by combining the asymptotic (surface) formulas of Lamb with the laws of reflection of plane waves and the reciprocal theorem, asymptotic expressions may be found for the interior displacements, for simple vibratory disturbances. Similar evaluations on a different basis are given by Miller and Pursey [1]. The step form of disturbance at an interior line source was investigated by Lapwood [1] for the semi-infinite medium with a free surface, and by Newlands [1] for an added solid surface layer.

The adaptation and development of Lamb's problem for predicting the interaction of a semi-infinite elastic foundation and a vibratory machine appears to begin with E. Reissner (1936). It has recently been carried further to include rocking and lateral motions, and interior evaluations by Bycroft [1], Miller and Pursey [1], and Arnold, Bycroft, and Warburton [1], who have obtained experimental substantiation.

* See also Blake, *J. Acoust. Soc. Amer.*, **24**, 211–215, 1952. Solutions for impulsive pressure are cited by Davies in *Appl. Mechanics Revs.*, **8**, p. 60, Review No. 356, 1955.

13. Concluding notes

Several important parts of the theory of elasticity have been omitted, in accordance with the statement of scope and intention in Section 1, even though there is substantial current research activity in them. One such is the *theory of finite deformations*. Beyond its early general theorems, there has been a baffling and growing variety of individual formulations of general theory leading to few tangible results which could be connected with technology or even with *ad hoc* experiment. These have been surveyed and clarified by Truesdell [1]. But within the past 10 years substantial results have sprung from the researches of Rivlin and the impulse these have given to the subject. They have been brought within a well-organized tensorial formulation by Green and Zerna [1], given in their *Theoretical Elasticity*, and their number is growing steadily. The literature of this development is in English, and nothing bearing on it has been encountered in the less accessible publications. It may be readily followed up, from Green and Zerna's book as introduction, through *Applied Mechanics Reviews* and *Mathematical Reviews*.* It is of primary interest not only as to rubber-like materials, but equally as to interaction (e.g., change of torsional stiffness due to tension) and stability in metallic members.

Other subjects have been omitted although there are more than a few little-known papers, mainly Russian, on them. These include small torsion and flexure; the stretching, bending, and twisting of bars which are slightly curved or twisted initially, or composite in structure—as a cylinder with longitudinal bores filled with different material; paraboloidal, ellipsoidal, and spherical boundaries; and general dynamical theory (rather than specific solutions) of elastic solids, especially as to vibrations. The Saint-Venant torsion problem each year acquires several new solutions by known methods, and naturally the degree of complexity tends to increase. The shapes involved for the most part exhibit the disadvantage emphasized in Section 2: they are mathematically attractive but otherwise improbable. The conformal representation of assignable shapes (Section 3) has less significance here than in the plane-stress plane-strain problems because of the ease with which torsion problems can be solved, in the practical sense, by finite-difference methods.

ACKNOWLEDGMENTS

The author's acknowledgments and thanks go to Marina Goodier for translations from the Russian, and to J. R. M. Radok and Brown University Library for loan of Russian books otherwise unobtainable.

* A survey by Doyle and Ericksen [1] appeared in 1956.

BIBLIOGRAPHY

Agababyan, E. H. [1], *Ukrain. Mat. Zhurn.*, **5**, 375–379, 1953.

Arnold, R. N., Bycroft, G. N., and Warburton, G. B. [1], *J. Appl. Mech.* **22**, 391–400, 1955.

Bogdanoff, J. L. [1], *J. Appl. Mech.*, **21**, 88, 1954.

Born, J. S., and Horvay, G. [1], *J. Appl. Mech.*, **22**, 401–406, 1955.

Burmistrov, E. F. [1], *Inzhen. Sbornik*, **17**, 199–202, 1953.

Bycroft, G. N. [1], *Phil. Trans. Roy. Soc. (London)*, A, **248**, 327–368, 1956.

Christie, D. G. [1], *Phil. Mag.* (7), **46**, 527–541, 1955.

Conway, H. D. [1], *J. Appl. Mech.*, **21**, 42–44, 1954.

Danilovskaya, V. I. [1], *Prikl. Mat. Mekh.*, **14**, 316–318, 1950.

——— [2], *Prikl. Mat. Mekh.*, **16**, 341–344, 1952.

Doyle, T. C., and Ericksen, J. L. [1], *Advances in Appl. Mech.*, **4**, 53–115, 1956.

Elliott, H. A. [1], *Proc. Cambridge Phil. Soc.*, **44**, 522, 1948.

Eshelby, J. D. [1], *Phil. Trans. Roy. Soc. (London)*, A, **877**, 87, 1951.

Eubanks, R. A., and Sternberg, E. [1], *J. Rat. Mech. Anal.*, **3**, 89, 1954.

Freudenthal, A. M., and Weiner, J. H. [1], *J. Appl. Phys.*, **27**, 44–50, 1956.

Fridman, M. M. [1], *Prikl. Mat. Mekh.*, **14**, 321–340, 1950.

——— [2], *Dokl. Akad. Nauk S.S.S.R.* (N.S.), **60**, 1145–1148, 1948.

——— [3], *Dokl. Akad. Nauk S.S.S.R.* (N.S.), **66**, 21–24, 1949.

Galin, L. A. [1], *Contact Problems of the Theory of Elasticity*, Moscow, 1953.

——— [2], *Dokl. Akad. Nauk S.S.S.R.* (N.S.), **39**, 1943.

Gatewood, B. E. [1], *Phil. Mag.* (7), **32**, 282–301, 1941.

Goodier, J. N., and Hsu, C. S. [1], *J. Appl. Mech.*, **21**, 147, 1954.

Green, A. E., and Zerna, W. [1], *Theoretical Elasticity*, Oxford, 1954.

Gutin, L. Ya. [1], *Akad. Nauk S.S.S.R. Zhur. Tekn. Fiz.*, **21**, 892–906, 1951.

Harding, J. W., and Sneddon, I. N. [1], *Proc. Cambridge Phil. Soc.*, **41**, 16, 1945.

Harvey, R. B. [1], *Proc. Roy. Soc. (London)*, A, **223**, 338–348, 1954.

Heller, S. R. [1], *J. Appl. Mech.*, **20**, 279–285, 1953.

Hieke, M. [1], *Z. angew. Math. Mech.*, **35**, 285–294, 1955.

——— [2], *Z. angew. Math. Mech.*, **35**, 54–62, 1955.

Hobson, E. W. [1], *Trans. Cambridge Math. Soc.*, **18**, 1900.

——— [2], *Proc. London Math. Soc.*, **22**, 1891.

Horvay, G. [1], *J. Appl. Mech.*, **20**, 87, 1953.

——— [2], *Proc. Second U. S. Nat. Congr. Appl. Mech.*, 1954.

Huth, J. H. [1], *J. Appl. Phys.* (U. S.), **23**, 1234–1237, 1952.

Isida, M. [1], *Proc. Jap. Nat. Congr. Appl. Mech.*, **2**, 1952.

Kikukawa, M. [1], *Proc. Jap. Nat. Congr. Appl. Mech.*, **4**, 1954.

Kikukawa, M. [2] *Proc. Jap. Nat. Congr. Appl. Mech.*, **3**, 1953.
—— [3], *Proc. Jap. Nat. Congr. Appl. Mech.*, **1**, 1951.
Kitover, K. A. [1], *Prikl. Mat. Mekh.*, **16**, 739–748, 1952.
Koiter, W. T. [1], *Quart. J. Mech. Appl. Math.*, **8**, 164, 1955.
Kosmodamianski, A. S. [1], *Prikl. Mat. Mekh.*, **16**, 249–252, 1952.
Kromm, A. [1], *Z. angew. Math. Mech.*, **28**, 104–114, and 297–303, 1948.
Kufarev, P. P. [1], *Dokl. Akad. Nauk S.S.S.R.* (N.S.), **23**, 1939.
—— [2], *Dokl. Akad. Nauk S.S.S.R.* (N.S.), **32**, 1941.
Kusukawa, Ken-ichi [1], *Proc. Jap. Nat. Congr. Appl. Mech.*, **2**, 1952.
Lamb, H. [1], *Phil. Trans. Roy. Soc. (London)*, A, **203**, 1–42, 1904.
Lapwood, R. E. [1], *Phil. Trans. Roy. Soc. (London)*, A, **242**, 1949.
Lekhnitzki, S. G. [1], *Anisotropic Plates*, 1947. (Russian.)
—— [2], *Theory of Elasticity of Anisotropic Bodies*, 1950. (Russian.)
—— [3], *Inzhen. Sbornik*, **19**, 83–106, 1954.
—— [4], *Prikl. Mat. Mekh.*, **4**, 1940.
Ling, Chih-Bing [1], *J. Appl. Mech.*, **19**, 141, 1952.
Lodge, A. S. [1], *Quart. J. Mech. Appl. Math.*, **8**, 211–225, 1955.
Lourie, A. I. [1], *Prikl. Mat. Mekh.*, **10**, 397–406, 1946.
—— [2], *Statics of Thin-Walled Elastic Shells*, 1948. (Russian.)
—— [3], *Inzhen. Sbornik*, **17**, 43–58, 1953.
—— [4], *Dokl. Akad. Nauk S.S.S.R.* (N.S.), **23**, 759–763, 1939.
—— [5], *Dokl. Akad. Nauk S.S.S.R.* (N.S.), **28**, 106–109, 1940.
—— [6], *Prikl. Mat. Mekh.*, **5**, 1941.
Lubkin, J. L. [1], *J. Appl. Mech.*, **18**, 183, 1951.
Maisel, W. M. [1], *Dokl. Akad. Nauk S.S.S.R.* (N.S.), **30**, 115–118, 1941.
Mansfield, E. H. [1], *Quart. J. Mech. Appl. Math.*, **6**, 370, 1953.
Maue, A. W. [1], *Z. angew. Math. Mech.*, **33**, 1, 1953.
—— [2], *Z. angew. Math. Mech.*, **34**, 1, 1954.
Melan, E. [1], *Österr. Ingen.-Arch.*, **8**, 165, 1954.
—— [2], *Ingen.-Arch.*, **20**, 46–48, 1952.
Mikhlin, S. G. [1], *Dokl. Akad. Nauk S.S.S.R.* (N.S.), **27**, 1940.
Miller, G. F., and Pursey, H. [1], *Proc. Roy. Soc. (London)*, A, **223**, 521–541, 1954.
Mindlin, R. D. [1], *Proc. Second U. S. Nat. Congr. Appl. Mech.*, 13–19, 1954.
Mossakowskii, V. I. [1], *Prikl. Mat. Mekh.*, **15**, 635–636, 1951.
—— [2], *Prikl. Mat. Mekh.*, **17**, 477–482, 1953.
—— [3], *Prikl. Mat. Mekh.*, **18**, 187–196, 1954.
Mossakowskii, V. I., and Zagubizenko, P. A. [1], *Dokl. Akad. Nauk S.S.S.R.* (N.S.), **94**, 409–412, 1954.
Mott, N. F. [1], *Proc. Roy. Soc. (London)*, A, **220**, 1–14, 1953.
Mura, T. [1], *Proc. Jap. Nat. Congr. Appl. Mech.*, **2**, 9–13, 1952.
Muskhelishvili, N. I. [1], *Some Basic Problems of the Mathematical Theory of Elasticity*, 3rd ed., Moscow-Leningrad, 1949, translated by J. R. M. Radok (Noordhoff, Groningen, 1953).
—— [2], *Singular Integral Equations*, 2nd ed., Moscow, 1946, translated by J. R. M. Radok (Noordhoff, Groningen, 1953).
Nabarro, F. R. N. [1], *Advances in Phys.*, **1**, 269–394, 1952.
Narishkina, E. [1], *Trudy Seism. Inst. S.S.S.R.*, No. 45, 1934.
Newlands, M. [1], *Phil. Trans. Roy. Soc. (London)*, A, **245**, 213, 1952.
Owens, A. J., and Smith, C. B. [1], *Quart. Appl. Math.*, **9**, 329–333, 1951.
Payne, L. E. [1], *Proc. Cambridge Phil. Soc.*, **50**, 466, 1954.

Petrashen, G. [1], *Dokl. Akad. Nauk S.S.S.R.* (N.S.), **64**, 649–652, 1949.
—————— [2], *Dokl. Akad. Nauk S.S.S.R.* (N.S.), **64**, 783–786, 1949.
Prokopov, V. K. [1], *Prikl. Mat. Mekh.*, **16**, 1949.
—————— [2], *Prikl. Mat. Mekh.*, **16**, 45–56, 1949.
Radok, J. R. M. [1], *J. Appl. Mech.*, **22**, 249–254, 1955.
Reissner, E., and Sagoci, H. [1], *J. Appl. Phys.*, **15**, 1944.
Reissner, H., and Morduchov, M. [1], *NACA Tech. Note 1850*, 1949.
Rosenfield, A. R., and Averbach, B. L. [1], *J. Appl. Phys.* (U. S.), **27**, 154, 1956.
Sadowsky, M. A. [1], *J. Appl. Mech.*, **22**, 177, 1955.
Saenz, A. W. [1], *J. Rat. Mech. Anal.*, **2**, 83–98, 1953.
Sauter, F. [1], *Z. angew. Math. Mech.*, **30**, 203–215, 1950.
Savin, G. N. [1], *Concentration of Stress around Holes*, Moscow-Leningrad, 1951. (Russian.)
—————— [2], *Dokl. Akad. Nauk S.S.S.R.* (N.S.), **23**, 1939.
—————— [3], *Dokl. Akad. Nauk S.S.S.R.* (N.S.), **24**, 1940.
—————— [4], *Vestnik Inzhenerov i. Tekh.*, 1940.
Selberg, H. L. [1], *Ark. Fysik*, **5**, 97–108, 1952.
Seremetev, M. P. [1], *Ukrain. Mat. Zhurn.*, **1**, 68–80, 1949.
—————— [2], *Inzhen. Sbornik*, **14**, 81–100, 1953.
—————— [3], *Ukrain. Mat. Zhurn.*, **5**, 58–79, 1953.
—————— [4], *Prikl. Mat. Mekh.*, **16**, 437–448, 1952.
Shapiro, G. S. [1], *Prikl. Mat. Mekh.*, **7**, 379–382, 1943.
Sherman, D. I. [1], *Prikl. Mat. Mekh.*, **15**, 297–316, 1951.
—————— [2], *Prikl. Mat. Mekh.*, **15**, 751–761, 1951.
—————— [3], *Izvest. Akad. Nauk S.S.S.R. Otdel. Tekh. Nauk*, **6**, 840–857, 1952.
—————— [4], *Trudy Seism. Inst. S.S.S.R.*, No. 118, 1946.
Shield, R. T. [1], *Proc. Cambridge Phil. Soc.*, **47**, 401, 1951.
Shtaerman, E. [1], *Dokl. Akad. Nauk S.S.S.R.* (N.S.), **29**, 182–184, 1940.
—————— [2], *Contact Problems of the Theory of Elasticity*, Moscow-Leningrad, 1949. (Russian.)
Smirnov, V. I., and Sobolev, S. L. [1], *Trudy Seism. Inst. S.S.S.R.*, No. 24, 1932.
Sneddon, I. N. [1], *R.C. Circ. Mat. Palermo* (2), **1**, 57–62, 1952.
Sobolev, S. L. [1], *Trudy Seism. Inst. S.S.S.R.*, No. 18, 1932.
—————— [2], *Mat. Sbornik*, **40**, 1933.
Sokolnikoff, I. S. [1], *Proc. Symposium Appl. Math.*, **3**, 1950.
Sveklo, V. A. [1], *Dokl. Akad. Nauk S.S.S.R.* (N.S.), **95**, 737–739, 1954.
Tarabasov, N. D. [1], *Inzhen. Sbornik*, **3**, 3–14, 1947.
—————— [2], *Dokl. Akad. Nauk S.S.S.R.* (N.S.), **63**, 15–18, 1948.
Truesdell, C. [1], *J. Rat. Mech. Anal.*, **1**, 125–300, 1952; **2**, 539–616, 1953.
Udoguchi, T. [1], *Proc. Jap. Nat. Congr. Appl. Mech.*, **4**, 1954.
—————— [2], *Proc. Jap. Nat. Congr. Appl. Mech.*, **3**, 1953.
Ugodchikov, A. G. [1], *Dokl. Akad. Nauk S.S.S.R.* (N.S.), **77**, 213–216, 1951.
Wells, A. A. [1], *Quart. J. Mech. Appl. Math.*, **3**, 23–31, 1950.
Williams, M. L. [1], *J. Appl. Mech.*, **19**, 526, 1952.
Yu, Yi-Yuan [1], *J. Appl. Mech.*, **19**, 537–542, 1952.
Zvolinski, N. V. [1], *Dokl. Akad. Nauk S.S.S.R.* (N.S.), **56**, 19–22, 1947.
—————— [2], *Dokl. Akad. Nauk S.S.S.R.* (N.S.), **59**, 1081–1084, 1948.
—————— [3], *Dokl. Akad. Nauk S.S.S.R.* (N.S.), **65**, 145–148, 1949.

The Mathematical Theory of
PLASTICITY

P. G. Hodge, Jr.
Illinois Institute of Technology

I NTRODUCTION

The word "plasticity" means many things to many people, and it is obvious that no single treatment can be compatible with all of these meanings. In view of the over-all purpose of this book, it has here been considered as a branch of applied mathematics. Thus, although the ultimate value of the theory is closely connected with the physical behavior of materials beyond the elastic limit, there are almost no specific references to experimental work in the field, nor has any attempt been made to justify the basic laws with reference to the crystalline or molecular structure of materials. Rather, certain basic postulates have been adopted and used as the starting point of a general theory.

In the course of this development, many topics which might have been included have not been considered at all. To some extent the choice of topics has been guided by the author's personal interests. However, it has also been guided by the current state of development of the field. Thus, for example, no mention has been made of plastic buckling. Despite the importance of this phenomenon to the engineer, and despite the numerous papers written upon the subject, there does not seem to be sufficient agreement among the experts to justify a unified approach to the subject. Other major topics which have been omitted for the same reason are plastic waves, time-dependent phenomena such as creep, and the extension of plasticity concepts to soil mechanics.

The history of plasticity theory reached somewhat of a milestone in 1950 and 1951 with the publication of four English language textbooks [0.1, 0.2, 0.3, 0.4],* the first to appear in 20 years. In fact, since Nadai's first book [0.5] in 1931, the only previous comprehensive treatment in

---

* Numbers in square brackets refer to the bibliography, which is listed by chapter and section at the end of this article.

any language was Sokolovsky's work in Russian, published in 1946 [24.23]. The present article is intended to be primarily a survey of the work done since that time. Therefore, space has not been devoted to topics which may be found in one or more of those four books unless it has been necessary to make the treatment reasonably self-contained.

When a material specimen is stressed in simple tension, its mechanical behavior is described by the stress-strain curve and a statement concerning its compressibility. Although the actual shape of the stress-strain curve depends upon the material being tested, certain features are common to almost all structural materials. For sufficiently small values of the stress, the relationship between stress and strain is linear and reversible. The behavior in this case is termed elastic. For somewhat higher stress values, the relationship becomes nonlinear and irreversible, and we have plastic behavior. The transition between the elastic and plastic behavior may be gradual as in Fig. 1a, or abrupt, as in Fig. 1b. In the former case, the yield stress σ^* at which the transition takes place is conventional, whereas in Fig. 1b it corresponds to a definite physical phenomenon.

The actual curves shown in Fig. 1a and b do not lend themselves readily to the solution of complex problems, so it is frequently desirable to approximate them by relatively simpler expressions. An obvious suggestion is to use a number of linear segments [0.6] and Fig. 1c shows a two-segment approximation to the curve of Fig. 1a. Segment AB corresponds to elastic behavior, and segment BC to plastic. If BC is horizontal, we obtain the curve of a perfectly plastic material shown in Fig. 1d. Finally, if the total strains are large compared to the elastic strains, the latter may reasonably be neglected, thus leading us to the theory of rigid-plastic solids (Fig. 1e and f).

Thus far, there is no way to distinguish between plasticity and nonlinear elasticity. This distinction appears when we consider unloading. The nonlinear elastic material will retrace its stress-strain curve to the origin, whereas the plastic process is irreversible. When the first increment of load is removed, the corresponding increment of strain will be elastic, so that the slope of the stress-strain curve will be the same as during the initial stages of loading. Thus, in Fig. 1, if the material is loaded into the plastic range along ABC and then unloaded, the initial slope of CD will be the same as that of AB.

If the unloading is continued and a negative load applied, then eventually the material will flow plastically in compression. For the perfectly plastic material the compressive yield stress will be a constant property of the material, and, for simplicity, we shall take it to be the same as the tensile yield stress. For a strain-hardening material, however, the compressive yield stress may depend upon the amount of tensile hardening which has taken place prior to unloading. This property will be discussed in more detail in Chapter 2.

Certain important differences between elasticity and plasticity can be seen immediately from the preceding discussion. Vital among these

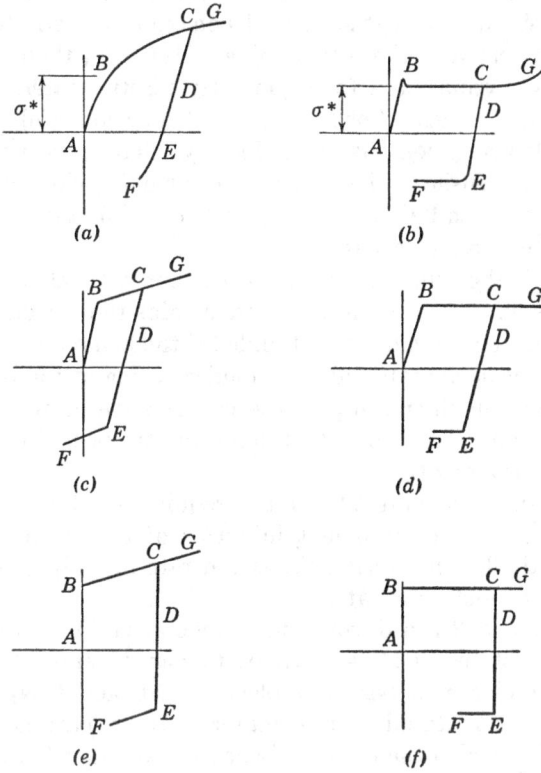

Fig. 1. Stress-strain curves. (a) Aluminum. (b) Mild steel. (c) Piecewise linear. (d) Perfectly plastic. (e) Rigid-linear hardening. (f) Rigid-perfectly plastic.

differences is the question of uniqueness. If the specimen is unloaded along path CDE in Fig. 1, the same states of stress and strain are traversed as in the original loading along ABC, but with a different correspondence. Thus the stress is no longer uniquely determined by the strain, and vice versa.

Not even the condition of being elastic or plastic can be determined from current information alone. Thus at a point D on the unloading path CDE in Fig. 1, the material is elastic, whereas at the strain corresponding to this point on the initial loading the point is plastic.

Since the resulting predictions are quite different for elastic or plastic behavior, it is important to formulate clearly the distinction between the two.

There are essentially two aspects to this distinction. On the one hand, after the material has gone through any given history of stress and strain two numbers, called the "current yield stresses" (tensile and compressive), will exist such that behavior is elastic so long as the stress remains between these two values. Initially, these two numbers are numerically equal, and for a perfectly plastic material they maintain their initial values. For a strain-hardening material, however, the values will depend upon the stress history. Thus, at point D in Fig. $1a$, c, or e the tensile and compressive current yield stresses are the stress levels of C and E respectively, whereas initially they were plus and minus the stress levels at B.

As stated, if the stress lies between the current yield stresses the behavior is elastic. But even if the stress attains the current yield stress, a second condition remains to be fulfilled if the behavior is to be plastic, namely that the stress rate must be nondecreasing at the tensile current yield stress and nonincreasing at the compressive current yield stress. Thus from point C we can either have plastic behavior along CG or elastic behavior along CD.

For the simple tension test this latter requirement for plastic flow may be given an alternate statement in terms of the strain rate, namely that for plastic flow in tension the strain rate must be positive, but in compression it must be negative.

Chapters 1 and 2 are devoted to a generalization of these concepts for multiaxial states of stress. In particular, Chapter 1 is concerned with the behavior of a perfectly plastic solid, and Chapter 2 with a strain-hardening material. In Chapter 3 some particular situations are discussed in which the material behavior exhibits a limited independence of the stress history. Then, in Chapter 4 we state and prove some of the extremum principles which are valid in plasticity theory.

The remainder of the survey is devoted to applications of the theory. Chapter 5 is a moderately exhaustive treatment of a particular elementary problem, chosen to illustrate the essential features of the various theories discussed. Then, in Chapters 6 and 7 we discuss other problems more briefly in an attempt to sketch the current state of development in plasticity problems. Particular attention is paid to significant Russian contributions.

CHAPTER 1

THEORY OF PERFECTLY PLASTIC SOLIDS

1. Generalized variables

If more than one component of stress is acting simultaneously, the description of material behavior becomes much more complex. In order to keep the ensuing discussion as general as possible, we shall follow Prager [1.1] and introduce generalized stress and strain variables. The stress variables will be denoted by $Q_1, Q_2, \cdots, Q_n$, where n is the number of variables needed to specify the stress state. For example, in a beam subjected to bending, the only variable would be $Q_1 = M$. In the torsion of a cylindrical rod we could set $Q_1 = \tau_{xz}, Q_2 = \tau_{yz}$. Alternatively, it might be convenient to use dimensionless variables $Q_1 = \tau_{xz}/k, Q_2 = \tau_{yz}/k$, where k is the yield stress in pure shear. The general three-dimensional stress state is included as a special case by setting $Q_1 = \sigma_x, Q_2 = \sigma_y, Q_3 = \sigma_z, Q_4 = \tau_{yz}, Q_5 = \tau_{zx}, Q_6 = \tau_{xy}$.

The choice of stress variables above is, to some extent, arbitrary. Thus in the torsion problem one might prefer to set $Q_1 = \tau_{rz}, Q_2 = \tau_{\theta z}$. Or, more profoundly, one might wish a more exact solution to the beam problem and set $Q_1 = \sigma_x, Q_2 = \sigma_y, Q_3 = \tau_{xy}$. However, once the stress variables have been chosen, the strain variables q_i are determined to within a single multiplicative constant by a requirement that the internal energy be of the form *

(1.1) $$U = C(Q_1 q_1 + Q_2 q_2 + \cdots + Q_n q_n) = C Q_i q_i$$

Thus, in the bending problem if $Q_1 = M$, then $q_1 = \kappa$. In the bending of circular plates which will be considered in more detail in Chapter 5,

---

* Throughout this paper Latin indices run from 1 to n, where n is the number of stress variables, and a repeated subscript is to be summed from 1 to n: $Q_i q_i \equiv \Sigma_{i=1}^{n} Q_i q_i$.

it is convenient to define

$$Q_1 = M_r/Yh^2, \qquad Q_2 = M_\theta/Yh^2$$

(1.2)

$$q_1 = a\kappa_r, \qquad q_2 = a\kappa_\theta$$

where Y is the tensile yield stress, $2h$ the plate thickness, and a the radius. In this case all stresses and strains are dimensionless and the internal energy per unit area is

$$U = M_r\kappa_r + M_\theta\kappa_\theta = (Yh^2/a)Q_iq_i$$

in agreement with (1.1).

A second example is provided by the radially symmetric deformation of axially symmetric shells. Let M_θ and M_ϕ be the moments per unit length, and N_θ and N_ϕ the corresponding direct stresses. We then define

(1.3a) $\qquad Q_1 = N_\theta/2Yh, \qquad Q_2 = N_\phi/2Yh, \qquad Q_3 = M_\theta/Yh^2$

$$Q_4 = M_\phi/Yh^2$$

as the stress variables, and

(1.3b) $\qquad q_1 = e_\theta, \qquad q_2 = e_\phi, \qquad q_3 = h\kappa_\theta/2, \qquad q_4 = h\kappa_\phi/2$

Here again it is readily verified that

$$U = 2YhQ_iq_i$$

In view of the defining relationship between stresses and strains, only quantities which contribute to the internal energy can be taken as generalized variables. Thus, in the usual beam theory the shear force S must be treated as a reaction, not as a stress, since shear deformations are neglected.

For linear elasticity, the generalized stresses and strains will be linear functions of the component stresses and strains respectively. It follows that they will be linearly related to each other, so that

(1.4) $\qquad\qquad\qquad q_i = B_{ij}Q_j$

The values of B_{ij} can always be computed in terms of Young's modulus and Poisson's ratio (or more generally in terms of the anisotropic elastic constants). Thus, for example, in the symmetric bending of circular plates, the elastic moments and curvatures are related by *

(1.5) $\quad M_r = \dfrac{2Eh^3}{3(1-\nu^2)}(\kappa_r + \nu\kappa_\theta), \qquad M_\theta = \dfrac{2Eh^3}{3(1-\nu^2)}(\kappa_\theta + \nu\kappa_r)$

---

* See, e.g., S. Timoshenko, _Theory of Plates and Shells_, McGraw-Hill, New York, 1940.

Solution of these equations for the curvatures, and substitution of the values from (1.2) leads to

$$(1.6) \qquad q_1 = (3aY/2hE)(Q_1 - \nu Q_2), \qquad q_2 = (3aY/2hE)(-\nu Q_1 + Q_2)$$

Therefore, in this case

$$(1.7) \qquad B_{ij} = \frac{3aY}{2hE}\begin{pmatrix} 1 & -\nu \\ -\nu & 1 \end{pmatrix}$$

In this example it is obvious that B_{ij} is symmetric. Also, since $0 \leq \nu \leq \frac{1}{2}$, it is positive definite. It can be shown that these two properties are generally true, and use will be made of this fact in Chapter 4.

A basic assumption of plasticity theory is that the total strain rate can always be decomposed into an elastic part and a plastic part. Thus, if $\dot{p}_i$ denotes the plastic strain rate, it follows from (1.4) that

$$(1.8) \qquad \dot{q}_i = B_{ij}\dot{Q}_j + \dot{p}_i$$

where dots indicate differentiation with respect to time. In much of the theory to follow we shall consider only the effect of the plastic strains p_i. If this is done, the results will be immediately meaningful only for a rigid-plastic material (Fig. 1e and f). However, by means of (1.8), any such results can be generalized to include elastic-plastic materials.

2. Yield condition and flow law

In the Introduction we showed that two current yield stress values were necessary to specify the elastic range of a one-dimensional stress system at any instant. This was because we had a choice of two directions of loading, tensile or compressive. When two or more independent generalized stress components exist at a point, an infinite number of possible loading directions exist, corresponding to any linear combination of the stress components. Therefore, there must be an infinite number of current yield stresses.

Since we cannot specify the elastic range by any finite number of values, we need a functional representation. Therefore, we shall assume that there exists some function f which depends generally upon the stress components and the previous stress history, such that when f is less than some preassigned number the material is elastic. For the particular case of a perfectly plastic material this yield function is, by definition, independent of the stress history and depends only upon the stress. Therefore, defining f in a suitably normalized fashion, we may write a condition for elastic behavior as

$$f(Q_1, Q_2, \cdots, Q_n) < 1$$

Further, it also follows from the definition of perfect plasticity that f can never be greater than 1.

In considering the remaining case $f = 1$, we must generalize the role played by the stress rate in simple tension. Since f does not change with stress history, it is evident that plastic flow can take place only if the value of f is and remains at unity. In other words, the function f itself cannot decrease for plastic flow to occur. The complete stress conditions for plastic and elastic behavior may now be stated as

(2.1a)   Plastic: $f = 1$ and $\dot{f} = 0$

(2.1b)   Elastic: $f < 1$ or $\dot{f} < 0$

These conditions may be conveniently visualized in a stress space whose co-ordinates are the variables Q_i. The equation $f = 1$ then represents a surface, and $f < 0$ is that side of the surface toward the

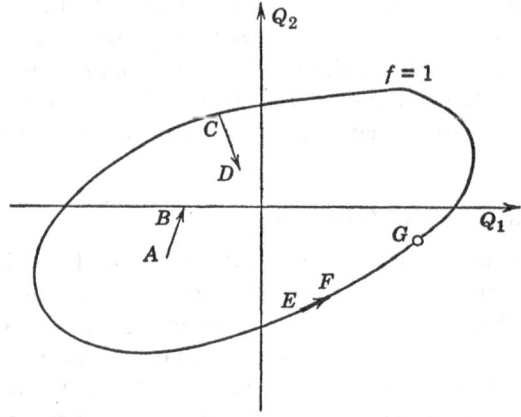

Fig. 2.   Elastic and plastic behavior of a perfectly plastic material.

origin. With reference to Fig. 2, (2.1) states that if the stress point moves from A to B or from C to D the behavior is elastic, whereas from E to F or remaining at G it is plastic.

Although in simple tension the condition on the strain rate is merely an alternative statement of the stress rate condition, it plays a much more vital role in multidimensional problems. In order to discuss this, we introduce the following postulate for the behavior of a perfectly plastic solid. Let a body be in an equilbrium state under an arbitrary set of body forces **f** and surface tractions **T**. Now, let some external agency apply an additional load to the body and then remove it. Then, the work done *by the external agency* during the loading is positive, and the work done *by the external agency* during the complete cycle of loading and unloading is nonnegative. This requirement was first formulated by Drucker [2.1, 2.2] and is essentially a statement of plastic irreversibility. In other words, energy put into a plastic deformation can never be recovered.

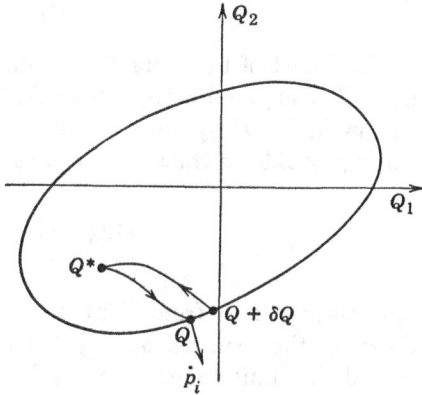

We shall discuss the consequences of this postulate in the geometrical terminology of Fig. 2. Suppose the equilibrium stress point under the loads **f** and **T** to be at Q^* (inside or on the yield surface) at a time $t = 0$. Let the external agency be such as to move it first to point Q on the yield surface at time t and then along the yield surface to point $Q + \delta Q$ at time $t + \delta t$. Removal of this agency then returns the

Fig. 3. Convexity of yield surface and normality of strain rate vector.

point to Q^* at a time t^* (see Fig. 3). Denoting the components of Q^* by Q_i^*, and the corresponding strains by q_i^*, etc., we may write the total work done during the cycle as

$$(2.2) \qquad \delta W_T = \int_0^t Q_i \dot{q}_i \, d\tau + \int_t^{t+\delta t} Q_i \dot{q}_i \, d\tau + \int_{t+\delta t}^{t^*} Q_i \dot{q}_i \, d\tau$$

Now, quite generally we may write the total strain rates as the sum of an elastic and plastic part:†

$$(2.3) \qquad \dot{q}_i = \dot{e}_i + \dot{p}_i$$

Then, since plastic straining can occur only from t to $t + \delta t$, (2.2) becomes

$$(2.4) \qquad \begin{aligned} \delta W_T &= \int_0^t Q_i \dot{e}_i \, d\tau + \int_t^{t+\delta t} Q_i (\dot{e}_i + \dot{p}_i) \, d\tau + \int_{t+\delta t}^{t^*} Q_i \dot{e}_i \, d\tau \\ &= \oint Q_i \dot{e}_i \, d\tau + \int_t^{t+\delta t} Q_i \dot{p}_i \, d\tau \end{aligned}$$

Here, the symbol $\oint$ refers to integration about the complete path, returning to the point Q^*. However, since by definition this integral

† The argument is not dependent upon linear elasticity, hence we use (2.3) rather than the more restrictive (1.8).

is elastic, the net work done in a closed cycle is zero, and (2.4) reduces
to

(2.5)
$$\delta W_T = \int_t^{t+\delta t} Q_i(\tau)\dot{p}_i(\tau)\,d\tau$$

Now, part of the work δW_T is done by the original equilibrium forces
Q_i^*. The expression for this work, δW_0 is obtained from (2.4), replacing
the variable Q_i by the constant Q_i^*. By an analysis similar to that
leading to 2.5 we then deduce that

(2.6)
$$\delta W_0 = \int_t^{t+\delta t} Q_i^*\dot{p}_i(\tau)\,d\tau$$

Subtracting (2.6) from (2.5) we are left with that portion of the work
done by the external agency. This quantity δW_e is then divided by
δt and the limit taken as δt tends to zero. The resulting expression is

(2.7)
$$\lim_{\delta t \to 0} \delta W_e/\delta t = [Q_i(t) - Q_i^*]\dot{p}_i$$

Therefore, the postulate that the external work be nonnegative during
the cycle is equivalent to the statement that

(2.8)
$$(Q_i - Q_i^*)\dot{p}_i \geq 0$$

Equation (2.8) imposes a severe restriction on the allowable shapes
of the yield surface $f = 1$. Geometrically, the equation states that the
vector from Q^* to Q must make an angle with the vector $\dot{p}_i$ of not
greater than $90°$, and this must be true for any Q^* in or on the yield
surface. Therefore, if a plane is drawn through Q perpendicular to $\dot{p}_i$,
then all admissible points Q^* must lie on or to one side of this plane.
This is precisely the definition of a convex surface, so that we have
proved the result, *the yield surface $f = 1$ must be convex.*

Next, let us take the convex yield surface as given and discuss the
restrictions on the plastic strain rate vector $\dot{p}_i$. Reversing the above
argument, $\dot{p}_i$ must make a nonobtuse angle with every vector $Q_i - Q_i^*$.
At a point on the surface with a uniquely defined normal, this can only
be satisfied if $\dot{p}_i$ is in the direction of the normal. Thus

(2.9a)
$$\dot{p}_i = \lambda\,\partial f/\partial Q_i$$

where λ is undetermined but nonnegative. At a corner or edge where
the normal is not unique, then the direction of $\dot{p}_i$ is not specified. However, if the unit normal approaches a finite number of linearly in-

dependent limiting values as the stress point approaches the singular point in question, then

$$(2.9b) \qquad \dot{p}_i = \Sigma \dot{\lambda}_\alpha \, \partial f_\alpha / \partial Q_i$$

where $\partial f_\alpha / \partial Q_i$ are the linearly independent gradients and each of the λ_α are nonnegative but otherwise arbitrary.

In conclusion, given the convex function f, the behavior of the perfectly plastic material is governed by (1.8) [or, more generally, (2.3)] relating elastic and plastic strains, (2.1) stating the condition for elastic or plastic behavior, and (2.9) giving the plastic flow law. Equation (2.9a) was first postulated without proof by Mises [2.3]. Other arguments deducing (2.9) have been given by Hill [2.4] and Thomas [2.5].

3. Definition of problems

The relations (2.9) obtained in the previous section are among the class of relations known as flow laws, since they express a relationship between the stresses and the strain rates. In general, they lead to mathematical problems which are difficult to solve, and various suggestions have been made for simplifying these problems. One such suggestion which has received considerable attention in the literature is to replace (2.9) by some finite relation between the stresses and strains.

As long as this substitution is made as a mathematical approximation, its validity can only be discussed in relation to the accuracy of its predictions in particular problems. However, many authors have gone one step further and claim that the deformation law relating stresses and strains is a legitimate physical relation. This statement is obviously false if unloading is to be considered, as was shown in the Introduction by the lack of a one-to-one correspondence between stress and strain even in simple tension. Therefore, proponents of deformation theory have usually limited their claims to programs which do not involve unloading.

Now, it can very easily be shown that even without unloading, (2.9) can predict two different strain values for the same stress. Thus, in Fig. 4 consider the loading path $OA'B$, where $A'BA''$ is a regular portion of the yield surface with a unique normal and nonzero curvature at each point. As the stress point moves along $A'B$, each increment of plastic strain must be normal to $A'B$ at that point, so that the resulting total strain p_i' must lie to the left of the normal n_i at B. On the other hand, if the stress point moves along $OA''B$ to the same final position B, the resulting total strain p_i'' must lie to the right of n_i. Therefore the total strain at B does depend upon the loading path and hence no deformation theory can yield the same result as (2.9). Exactly the same arguments can be applied to a strain-hardening material or to a material whose yield function f depends upon other quantities such as strain or stress rate in addition to stress.

Before leaving this topic it should be noticed that the preceding demonstration is not applicable if the segment $A'BA''$ is plane. We

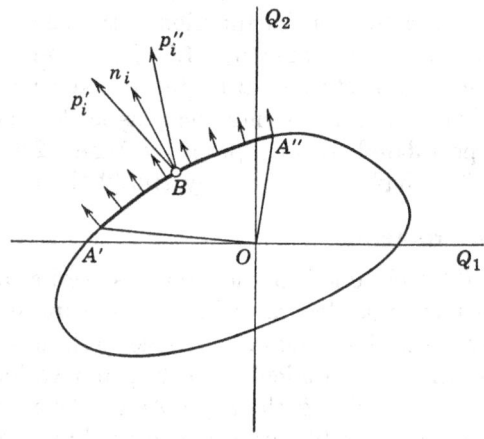

Fig. 4. Dependence of final strain on loading path.

shall see in Chapter 3 that in this case a certain type of limited deformation law can indeed be utilized. Other particular cases of agreement have been investigated by Edelman [3.1] and Handelman and Warner [3.2].

Since the relation between stress and strain is not unique, it is obviously absurd to formulate a boundary value problem in terms of instantaneous stresses and strains only. Instead, we must prescribe suitable combinations of boundary displacements and tractions and body forces, all as functions of time starting from the stress-free state. From the mathematical viewpoint it is evident that this is equivalent to prescribing an existing state of stress and strain throughout the body, together with traction rates and velocities on the boundary and body force rates throughout. A precise formulation of this problem, together with a discussion of the uniqueness of a solution, will be given in Section 11.

CHAPTER 2

THEORY OF STRAIN-HARDENING PLASTIC SOLIDS

4. Yield condition and flow law

The elastic behavior of a strain-hardening solid is the same as that of a perfectly plastic one. Therefore, it follows that the conditions for initial yield must be the same. Indeed, the difference between the two concerns only the mechanism for continuing plastic flow, plus the fact that the conditions for current yielding will depend upon the plastic history of the material. In stating the conditions for continuing plastic flow, then, we must replace (2.1) by

$$(4.1a) \qquad \text{Plastic:} \ f = f_c \ \text{and} \ \dot{f} > 0$$

$$(4.1b) \qquad \text{Elastic:} \ f < f_c \ \text{or} \ \dot{f} \leq 0$$

Here the function f depends upon the plastic stress history and $f = f_c$ represents the current yield condition.

Now, the remaining discussion of Section 2 made no specific use of (2.1), and it follows that exactly the same results are valid based on (4.1). Indeed, whereas in Section 2 (2.8) is expressed in an "equal to or greater than" relation, it can easily be shown that for the strain-hardening material the equality sign can hold only if $\dot{p}_i = 0$ [2.2]. In particular, then, at any stage of loading the current yield function must be convex, and the plastic strain rate vector must be normal to this surface.

However, although we may still apply the same restrictions to the current yield function at any time, we do not as yet have any such well-defined criteria as to the manner in which this function depends on the plastic loading history. Such experimental evidence as is available [4.1, 4.2, 4.3] indicates that the situation is extremely complex. This circumstance, together with the fact that there is very little significant experimental work available on the subject, has left the field wide open for theoretical assumptions.

Even in the case of simple tension and compression the situation is not clear. When the specimen is unloaded, after reaching a stress $\sigma_1 > \sigma^*$, the yield point for compressive yield will vary for different materials and under different conditions. Approximately at one extreme, the elastic unloading range will be double the initial yield stress, so that the material will yield in compression when

$$(4.2) \qquad \sigma = \sigma_1 - 2\sigma^*$$

This is shown as path $ABCDE$ in Fig. 5. Thus, according to this theory, the total elastic range of the material remains constant, but the actual

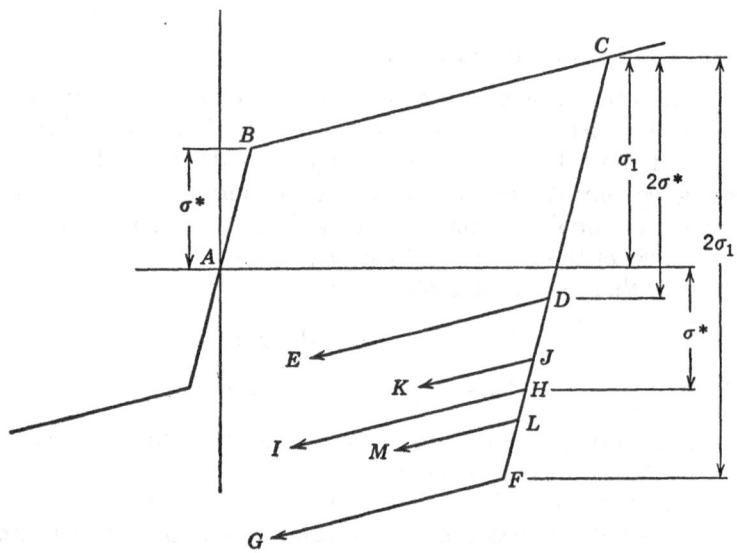

Fig. 5. Theories of reverse loading. *DE:* Bauschinger. *FG:* Isotropic. *HI:* Fixed. *JK, LM:* General.

compressive yield stress is reduced in magnitude due to the tensile yielding.

At the other exteme lies isotropic hardening. According to this theory, the mechanism that produces hardening acts equally in tension and compression, even though the yielding is in tension. Thus compressive yielding starts when

$$(4.3) \qquad \sigma = -\sigma_1$$

as shown by path $ABCFG$ in Fig. 5. In this case, the compressive yield stress and the elastic range are both numerically increased by the hardening.

Halfway between these two extremes is a theory based upon the independence of compressive and tensile hardening, according to which the compressive yield stress remains fixed at

$$(4.4) \qquad\qquad \sigma = -\sigma^*$$

independently of the amount of hardening (path $ABCHI$ in Fig. 5). This case, as well as any other criterion between (4.2) and (4.3) (paths $ABCJK$ and $ABCLM$ in Fig. 5) may be obtained as linear combinations of the two extreme cases.

In the remainder of this chapter we shall consider appropriate generalizations of these various behaviors. It should be mentioned at the outset that much of the material to be presented here is of recent origin, and has not been adequately tested with regard to its relationship to real physical materials. However, the theories are mathematically consistent and appear to offer a reasonable probability of leading to significant results in the near future.

5. Kinematic hardening

As our first hardening law we shall consider a generalization which maintains the elastic range. Although the basic concept involved is easily extended to an arbitrary hardening curve, we shall here restrict ourselves to piecewise linear rigid hardening (Fig. 1e). As mentioned in Section 1, any results we obtain can be immediately generalized to a linear elastic-hardening material by means of (1.8).

With these restrictions, there results a most ingenious kinematic model, first used by Prager in 1955 [5.1]. Considering first the simple tension-compression test, we adopt as our model a slotted bar as indicated in Fig. 6a. The bar is free to move along its length on the table TT, but will do so only when it is engaged by the pin P. Initially the pin P is at the center R of the slot and this point is marked as point O on the table. The distance from P to either end of the slot is taken equal to the yield stress σ^* of the rigid-linear hardening material.

If the pin P is now moved in such a way that the distance OP is always equal to the stress, the kinematic model will furnish the strain at any time and the conditions for further plastic flow. The strain will be proportional to the distance OR, and will, in fact, be given by

$$\epsilon = \overline{OR}/c$$

where $\tan^{-1} c$ is the slope of the plastic stress-strain curve. Thus, plastic flow will take place only when the pin is engaged at one end or the other of the slot. Figures 6 and 7 show a typical stress-strain history in terms of the kinematic model and stress-strain diagram respectively. The points marked A, B, C, etc., in Fig. 7 correspond to the positions (a), (b), (c), etc., of the model in Fig. 6.

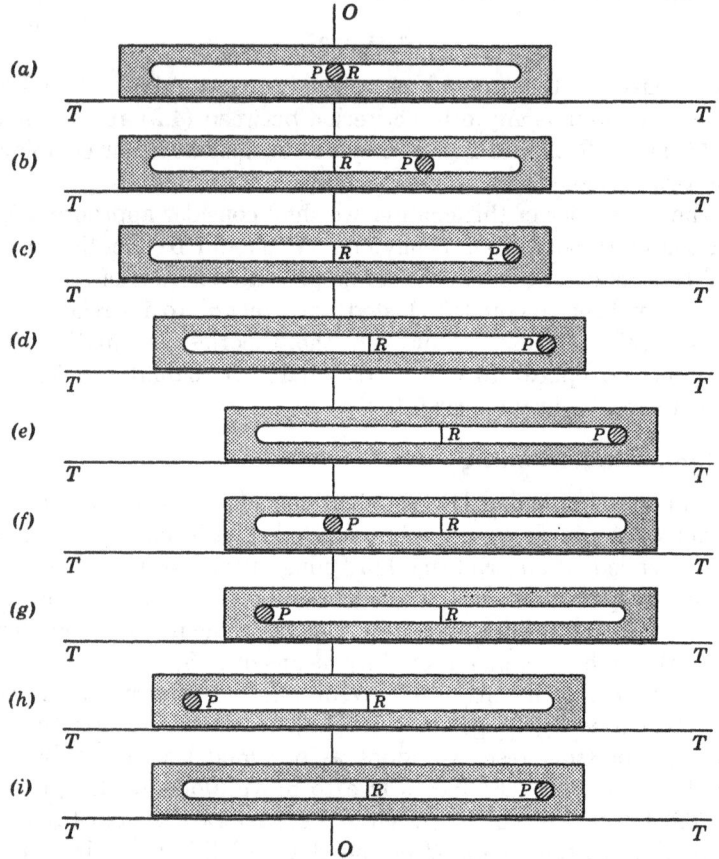

Fig. 6. Kinematic model for rigid-linear hardening tension-compression test. $\sigma = \overline{OP}$, $\epsilon = \overline{OR}/c$.

In a biaxial state of stress, two stress variables Q_1 and Q_2 may be specified, as opposed to the single variable σ. Therefore, if the position of the pin is to indicate the state of stress, it must be free to move in a two-dimensional space. In such a space, the rigid region must be the interior of a frame bounded by the initial yield condition, rather than a slot bounded by $\pm\sigma^*$. Thus we are led to the kinematic model shown in Fig. 8a. The actual shape of the frame will depend upon the initial yield function of the material. Most generally, it will consist of a combination of curved sides, straight sides, and corners. However, for simplicity we shall consider only the situation where all sides are straight. The stress values Q_1 and Q_2 are the co-ordinates of point P; the coordinates of R are given by cp_1 and cp_2, p_i being the plastic strain components.

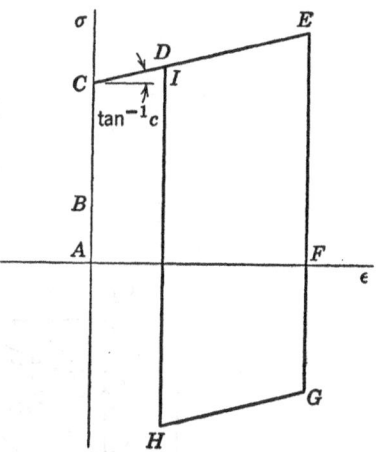

Figure 8a shows a typical history. Initially pin P and frame center R are at the origin. When P has reached the point P_1 it is still in the interior of the frame. Therefore, the frame has not moved and R is still at the origin. At P_2 the stress point engages the side of the frame. The frame is assumed to be perfectly smooth and to be constrained against rotation, so that its motion will be influenced only by the Q_1 component of the motion of P. Thus, from P_2 to P_3 the frame center moves along the Q_1 axis to R_3.

Fig. 7. Rigid-linear hardening stress-strain diagram.

At P_3, the stress point engages the corner of the frame and stays there until P_4, the frame center moving from R_3 to R_4. Then the direction of motion changes so that the stress point slides out of the corner but is still in contact with the top side. This continues through P_5 to P_6 when a different corner is engaged. Then at P_7 the direction of motion is changed and the material is unloaded to P_8, the frame center staying at R_7. At P_8 a different corner is engaged and plastic behavior resumes.

Two interesting features of the theory may be deduced directly from the model. In the first place, it is evident that proportional loading does not necessarily produce proportional strains. Thus, the straight load path from O to P_4 represents proportional loading, but the strain q_2 is zero until P_3 and then increases, whereas q_1 starts increasing at P_2.

A second feature is the fact that the final state is partially independent of the load path. Thus in Fig. 8b, if the stress point proceeds directly from the origin to P_1, the frame center will be at R_1. On the other hand, if the stress point moves along $OP_2P_3P_1$, the frame center will again end up at R_1. However, if the stress point moves along OP_4P_1, the frame center moves to the quite different location of R_1'. It should

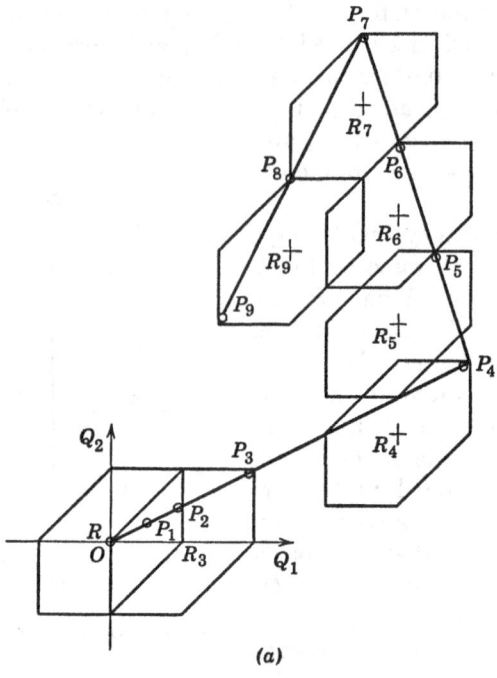

(a)

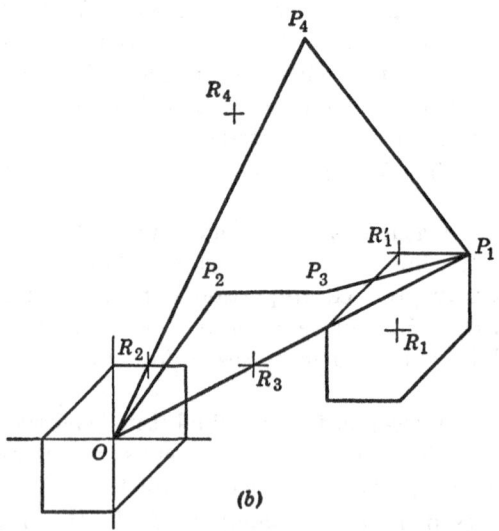

(b)

Fig. 8. (a) Kinematic model for biaxial stress. Co-ordinates of P are Q_1, Q_2; co-ordinates of R are cp_1, cp_2. (b) Partial independence of final state from loading path.

be noted that this has nothing to do with the concepts of loading or unloading, since the pin remains in contact with the frame along each of the three loading paths. Further consequences of this partial independence will be discussed in the next chapter.

Up to now we have spoken about two stress variables, Q_1 and Q_2. However, the generalization to n stress variables $Q_1, Q_2, \cdots, Q_n$ is obvious. We merely need to consider the yield frame as an $(n-1)$-dimensional manifold in an n-dimensional stress space.

When one or more of the generalized stress variables used to define a problem vanish, a distinction must frequently be made between two methods of applying the model of kinematic hardening [5.2, 5.3]. Simple kinematic hardening suppresses the zero stress component from the start and sets up the yield frame in a reduced number of dimensions. Complete kinematic hardening, on the other hand, considers all stresses, including any which happen to be zero.

The difference in prediction between the two theories is illustrated in Fig. 9. Let the initial yield frame in terms of Q_1 and Q_2 be the rec-

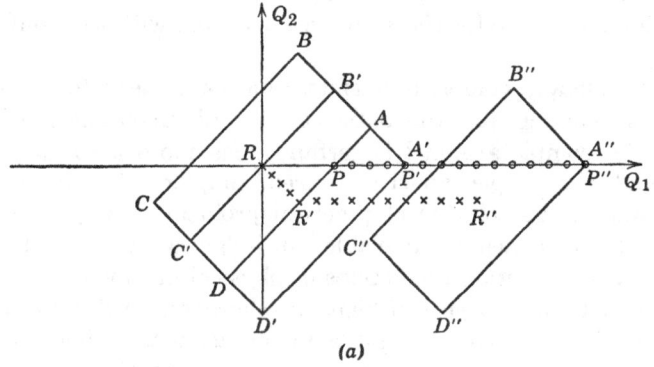

(a)

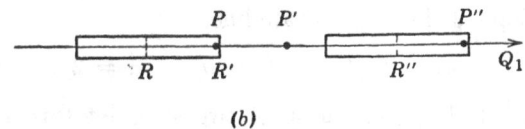

(b)

Fig. 9. Simple (b) and complete (a) kinematic hardening. In (a) the Q_2 effect is included; in (b) it is neglected.

tangle $ABCD$, and consider a problem where Q_2 is always zero. A yield frame similar to $ABCD$ would be used to represent the combined bending and tension of an idealized I-beam with unequal flanges. In this case Q_1 represents the bending moment and Q_2 the axial force, and we are considering a problem where the axial force is zero.

As Q_1 is increased, the pin in Fig. 9a first engages the frame on side AD. While the stress point moves from P to P', the frame moves normal to side AD, so that the center moves from R to R'. Thus the strain rate during this stage is

(5.1) $$c\dot{q}_1 = -c\dot{q}_2 = \tfrac{1}{2}\dot{Q}_1$$

At point P' the pin engages the corner A, and from there on the frame center moves in the same direction as the pin. Thus

(5.2) $$c\dot{q}_1 = \dot{Q}_1, \qquad c\dot{q}_2 = 0$$

Now, if Q_2 were completely neglected, the one-dimensional frame shown in Fig. 9b would be used. In this case, the strain rates would be predicted by (5.2) for the entire path $PP'P''$.

A comparison of (5.1) and (5.2) shows that the simple kinematic model fails to predict the change in strain q_2 which is associated with the complete description. More important than this, however, it predicts a different value for the strain q_1 associated with the nonvanishing stress Q_1.

Thus simple and complete kinematic hardening lead to two different results. Assuming that the basic concept of kinematic hardening is correct, the complete model is certainly the more accurate. On the other hand, the simple model is, indeed, simpler. As such it may be worth using in the solution of practical problems. As yet insufficient experimental evidence is available to judge whether or not either model provides accurate predictions of physical phenomena.

The geometrical concepts of kinematic hardening will now be formulated analytically. When the pin is in contact with a single side of the frame, the strain rate vector must be normal to that side. Let the stress axes be rotated so that the side in question is perpendicular to the N_1 axis and contains the remaining axes $T_2, T_3, \cdots, T_n$. Then, if c is the slope of the stress-strain line,

(5.3) $$\dot{n}_1 = \dot{N}_1/c, \qquad \dot{t}_\alpha = 0, \qquad \alpha = 2, 3, \cdots, n$$

Similarly, at the intersection of two sides let this intersection be perpendicular to the N_1 and N_2 axes and contain the remaining $T_3, T_4, \cdots, T_n$ axes. Then

(5.4) $$c\dot{n}_1 = \dot{N}_1, \qquad c\dot{n}_2 = \dot{N}_2, \qquad \dot{t}_\alpha = 0, \qquad \alpha = 2, 3, \cdots, n$$

The generalization to a p-fold intersection is obvious and leads to

(5.5) $$c\dot{n}_\alpha = \dot{N}_\alpha(\alpha = 1, 2, \cdots, p), \qquad \dot{t}_\alpha = 0(\alpha = p + 1, p + 2, \cdots, n)$$

Before leaving this topic, it is interesting to note that the kinematic model can also be applied to a perfectly plastic material. To this end, we merely define the plastic strain components as directly equal to the co-ordinates of R, and the stress components as the components of a vector drawn from the frame center to the pin, rather than from the origin to the pin. The validity of these definitions may be left as an exercise.

6. Isotropic hardening

Whereas the kinematic model theory assumes that the initial yield surface retains its size, shape, and orientation, but is free to translate as a rigid body, isotropic theory assumes that the yield surface maintains its shape, center, and orientation, but expands uniformly about the origin. Figure 10 shows the predicted yield surfaces according to

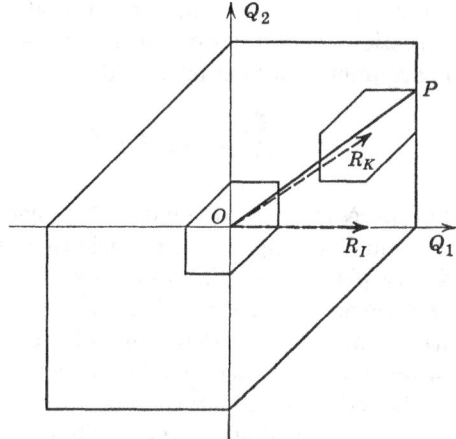

Fig. 10. Kinematic and isotropic theories for same loading path OP.

the two theories at the end of the same loading path OP. Not only are the yield surfaces for further yielding different, but so are the strains as indicated by the points R_K and R_I.

Although we have been able to show the isotropic strain point for a particular type of loading in Fig. 10, there does not seem to be any general simple method of incorporating it into such a model. Therefore, we shall proceed directly to the mathematical formulation of the flow law. We first recall that the initial yield surface is $f = 1$, and that by definition this surface expands uniformly about the origin. Therefore, the equation of the current yield surface at any time is

(6.1) $$f = f_{\max}$$

where f_{max} is the maximum value (> 1) previously attained by f. Therefore, (4.1) may be specified in the form

(6.2a)   Plastic: $f = f_{max}$ and $f \geq 1$ and $\dot{f} > 0$

(6.2b)   Elastic: $f < f_{max}$ or $f < 1$ or $\dot{f} \leq 0$

Assuming for the moment that the level surface of f has a unique normal, it follows from the arguments of Sections 2 and 4 that the plastic strain rate components are directed along this normal. Further, since all time effects such as inertia or creep have been neglected, we can write the plastic flow law in the form

(6.3) $$\dot{p}_i = F(\partial f / \partial Q_i)\dot{f}$$

where the yield function f and the hardening function F may both depend upon the stress point Q_i.

If the surface f does not have a unique normal at a point P, let f_1, $f_2, \cdots, f_p$ be the equations of the various surfaces meeting at P. Then it follows from the arguments of Section 2 that

(6.4) $$\dot{p}_i = F \sum_{\alpha=1}^{p} \lambda_\alpha (\partial f_\alpha / \partial Q_i)\dot{f}$$

Note that it is not necessary to distinguish between the various $\dot{f}_\alpha$, because they are all equal so long as the particle remains in the corner.

Now, for a perfectly plastic material all of the λ_α are arbitrary positive quantities. However, for a strain-hardening material the overall length of the strain rate must be determined by the hardening function, so that there must exist a further relation between the λ_α. This relation may be obtained from the postulate that the function $\dot{Q}_i p_i$ depends only upon the stress and stress rate. Thus, since

(6.5) $$\dot{Q}_i \dot{p}_i = F \sum_{\alpha=1}^{p} \lambda_\alpha (\partial f_\alpha / \partial Q_i)\dot{Q}_i \dot{f} = F \left(\sum_{\alpha=1}^{p} \lambda_\alpha \right) \dot{f}^2$$

then $\Sigma \lambda_\alpha$ must be equal to a constant. Finally, a consideration of the special case where $p = 1$ shows that

(6.6) $$\sum_{\alpha=1}^{p} \lambda_\alpha = 1$$

Equation (6.6) may also be obtained by various limiting processes for the corner [6.1, 6.2].

The isotropic and kinematic model theories may be linearly combined to furnish a possible generalization of any of the reloading curves of

Fig. 5 [6.3]. However, since no application of such a general combination has been made as yet, we shall postpone this topic and consider a special case of it in the next chapter. Another similar generalization without any restriction to linear hardening has been proposed by Besseling [6.4].

7. Other types of hardening

The theories discussed in Sections 5 and 6 are not by any means the most general generalizations of the various reverse loading behaviors illustrated in Fig. 5. Thus, for example, the hardening rate c may be different in different directions. Such anisotropic hardening may exist

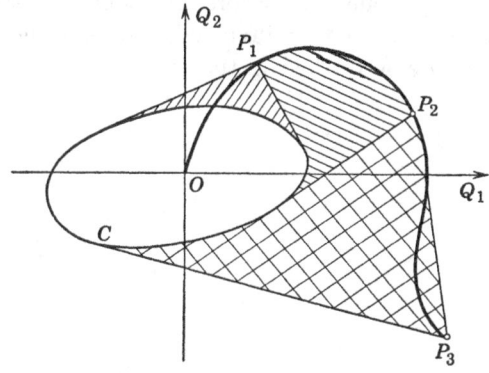

Fig. 11.   Growth of yield function according to slip theory.

conjointly with or independently of anisotropic yielding. Both the kinematic model and isotropic models apply immediately to the latter phenomenon, since they are independent of any assumed symmetries of the yield function. Both theories can also be generalized to include anisotropic hardening rates.

All of the theories discussed so far assume that the shape of the yield curve remains constant, although it may change in size or location. This restriction is based on mathematical convenience, rather than upon any physical or mathematical principle, and some experimental evidence is available [7.1, 7.2] that it is not particularly valid. A more general piecewise linear theory which includes a change in shape has been proposed in [5.3].

Starting with the physical concept of grain slip, Batdorf and Budiansky [7.3] have derived a theory which predicts quite a different type of growth of the yield function. Thus, if the initial yield function is some curve C and the stress point follows the curve OP, then the current yield function is the convex hull of C and OP. In Fig. 11 the unshaded region is the initial elastic region and the various shaded regions indicate additional elastic regions corresponding to P_1, P_2, and P_3 respectively. Note that the elastic region always adds to itself, never subtracts.

One feature of this so-called "slip" theory is that the stress point is almost always at a corner. The resulting flow laws are more complex in this case, and lead to equations which are not easily handled. Partly for this reason comparatively few problems have been solved for slip theory. Another reason is that some experimental results [7.4, 7.5] indicate rather less than perfect agreement. If one must be content with imperfect results there appears little reason for preferring a complex theory to a simple one. Therefore, although it is still too early to fully assess the relative merits involved, we shall not attempt any more complete discussion of the slip theory.

CHAPTER 3

PIECEWISE LINEAR PLASTICITY

8. Perfectly plastic solids

The analysis presented in Chapter 1 is valid for any shape of yield surface. However, when it comes to the specific solution of problems, a complex yield surface introduces tremendous difficulties. Therefore, although the experimental work of Stockton and Drucker [4.1] and Phillips [4.2, 4.3] indicates that the actual yield surface is complex, some approximations must be made for the sake of obtaining any solutions. In the present chapter we shall investigate the effect of a piecewise linear approximation to this surface. This concept was first used by Koiter [8.1] in 1953 to investigate the stresses and strains in a thick-walled tube under pressure. Other solutions to specific problems were obtained by Prager [8.2] and Hodge [8.3], but the first systematic development of the theory was begun by Prager [5.1] in 1955.

We assume that the yield surface consists of a finite number m of faces whose equations may be written

$$(8.1) \qquad f = a_{\beta i} Q_i = 1, \qquad \beta = 1, 2, \cdots, m$$

A considerable amount of flexibility is present in choosing the approximation to any given yield surface, since a suitable choice with m large can render the approximation very accurate, whereas with fewer faces the resulting analysis will generally be simpler.

If the stress point is in contact with the single face

$$(8.2) \qquad f = a_{1i} Q_i = 1$$

we say that it is in plastic regime 1. It follows from (1.8) and (2.9a) that the flow rule for this case is

$$(8.3) \qquad \dot{q}_i = B_{ij} \dot{Q}_j + \lambda a_{1i}$$

Now, let us assume that the stress point moved directly from the virgin elastic state to regime 1 at a time t_0. Then (8.3) may be integrated to yield

$$(8.4) \qquad q_i = B_{ij}Q_j + \lambda a_{1i}$$

Since $\dot{\lambda}$ is nonnegative, so is λ. Also, continuity with the elastic strains at $t = t_0$ is obtained by assigning $\lambda(t_0) = 0$.

Next, let us consider a particle which moves from the elastic state to face 1 at time $t = t_0$, and then to the intersection of faces 1 and 2, regime 1-2, at time $t = t_1$. For $t > t_1$ it is assumed to remain at the intersection, so that

$$(8.5) \qquad a_{1i}Q_i = a_{2i}Q_i = 1$$

The appropriate flow law is obtained from (2.9b) and may be written

$$(8.6) \qquad \dot{q}_i = B_{ij}\dot{Q}_j + \dot{\lambda}a_{1i} + \dot{\mu}a_{2i}, \qquad t_1 < t$$

For $t_0 < t < t_1$, (8.4) holds, so that the proper integral of (8.6) is

$$(8.7) \qquad q_i = B_{ij}Q_j + \lambda a_{1i} + \mu a_{2i}, \qquad t_1 < t$$

where $\mu(t_1) = 0$. It will be observed that the time t_1 at which the transition takes place does not explicitly enter (8.7).

Generally, a stress point will be said to "progress regularly" if it goes from the elastic state to face 1, say, then to the intersection of faces 1 and 2, then to the multiple intersection 1, 2, 3, etc. In other words, a regularly progressing stress point never loses contact with a face of the yield surface. The time intervals in each step need not be nonzero, however, so that a stress point may move directly from regime 1 to regime 1-2-3, for example. Obviously the correct generalization of the integrated flow law for the interval spent in a p-multiple intersection is

$$(8.8) \quad q_i = B_{ij}Q_j + \sum_{\beta=1}^{p} \lambda_\beta a_{\beta i}, \qquad Q_i a_{\beta i} = 1, \qquad \beta = 1, 2, \cdots, p$$

If the stress point does not stay in the intersection of 1 and 2 but moves on to face 2, the situation is not quite so simple. In this case

$$(8.9) \qquad \dot{q}_i = B_{ij}\dot{Q}_j + \dot{\mu}a_{2i}, \qquad t_1 < t$$

hence

$$(8.10) \qquad q_i = B_{ij}Q_j + \mu a_{2i} + \lambda(t_1)a_{1i}$$

where $\mu(t_1) = 0$. Thus, in this situation the strains do depend upon the time of transition.

Although the stress-strain relation (8.4) is apparently solved for the strain, this is true only to within the unknown quantity λ. However, if the state of strain is given, then (8.2) and (8.4) can be solved for the stresses in terms of the strains, for an elastic-plastic material. To this end, let b_{ij} denote the inverse matrix of B_{ij} so that

$$b_{ki}B_{ij} = \delta_{kj}$$

Since the elastic relationship is known to be reversible (for $0 < \nu < \frac{1}{2}$), b_{ik} exists and is symmetric. Multiplying (8.4) by b_{ki}, we obtain

(8.11) $$b_{ki}q_i = Q_k + b_{ki}a_{1i}\lambda$$

Next, multiplication of (8.11) by a_{1k} and use of (8.2) leads to

$$b_{ki}a_{1k}q_i = 1 + b_{ki}a_{1i}a_{1k}\lambda$$

or

(8.12) $$\lambda = \frac{b_{ki}a_{1k}q_i - 1}{b_{ki}a_{1i}a_{1k}}$$

Therefore, with λ known, (8.11) may be solved for the stresses:

(8.13) $$Q_k = b_{ki}q_i - b_{ki}a_{1i}\frac{b_{rs}a_{1r}q_s - 1}{b_{rs}a_{1r}a_{1s}}$$

Similar results can be obtained for the general multiple intersection (8.8). Also, it can be shown that the requirements that λ be positive and that the stress point not lie outside any other face uniquely determine the proper face or faces with which the stress point is in contact. However, it must be pointed out that this inversion is possible only for the elastic-plastic material and becomes meaningless in the rigid-plastic case. This indicates that in problems where the elastic strains are small compared to the plastic ones (8.13) may become unduly sensitive to small variations.

9. Strain-hardening solids

If the yield function is piecewise linear, then the results of Sections 5 and 6 can also be expressed in finite form, provided that the stress point progresses regularly. For simplicity, it will be assumed in the following that the hardening function is constant, although, as Sanders [9.1] has shown, more general hardening functions can be integrated.

We consider first the case of isotropic hardening [6.1, 6.2, 9.2], and assume the yield surface to consist of a finite number of faces

(9.1) $$f_\alpha = a_{\alpha i}Q_i$$

Then, if the stress point is on a single face (6.3) may be written *

$$\dot{p}_i = Fa_{\alpha i}\dot{f} = Fa_{\alpha i}a_{\alpha j}\dot{Q}_j$$

and the total strain rate is

(9.2) $$\dot{q}_i = B_{ij}\dot{Q}_j + Fa_{\alpha i}a_{\alpha j}\dot{Q}_j$$

where F is constant. Similarly, at a multiple intersection

(9.3) $$\dot{q}_i = B_{ij}\dot{Q}_j + F\sum_{\alpha=1}^{p}\lambda_\alpha a_{\alpha i}\dot{f}, \qquad \sum_{\alpha=1}^{p}\lambda_\alpha = 1$$

If the stress point moves directly from the elastic regime to face 1, then the correct integral of (9.2) is

(9.4) $$q_i = B_{ij}Q_j + Fa_{\alpha i}(a_{\alpha j}Q_j - 1)$$

With the stated restriction on the loading path, (9.4) is a one-to-one relation between the total stress and total strain. Similarly, if the stress point moves in a regular progression to a multiple corner, (9.3) can be integrated to yield

(9.5) $$q_i = B_{ij}Q_j + F\int_{1}^{f}\sum_{\alpha=1}^{p}\lambda_\alpha a_{\alpha i}\,df, \qquad \sum_{\alpha=1}^{p}\lambda_\alpha = 1$$

where the λ_α are defined so that $\lambda_\beta = 0$ up until face β becomes part of the intersection.

Similar results are available for the kinematic model hardening material and for the generalized hardening mentioned at the end of Section 6. However, in order to present the basic ideas in their simplest form, we shall consider a particular example, rather than carry out the most general discussion. To this end, let the discussion be confined to the two faces

(9.6) $$f_1 = Q_1, \qquad f_2 = Q_2$$

in a two-dimensional stress space. Further, we shall neglect elastic strains and consider a rigid-plastic material.

With these simplifications, (9.4) and (9.5) for an isotropic material show that

(9.7a) $$q_1 = F(Q_1 - 1), \qquad q_2 = 0, \qquad \text{on face 1}$$

(9.7b) $$q_1 = 0, \qquad q_2 = F(Q_2 - 1), \qquad \text{on face 2}$$

---

* The summation convention does not apply to Greek subscripts, since these are not co-ordinates.

$(9.7c)$ $\qquad q_1 = F \int_1^f \lambda_1 \, df, \qquad q_2 = F \int_1^f \lambda_2 \, df \qquad$ on corner 1-2

Now, since $\lambda_1 + \lambda_2 = 1$, the two Eqs. $(9.7c)$ may be added to eliminate λ_1 and λ_2. Recalling that $Q_1 = Q_2$ at the corner, we obtain

$(9.7d)$ $\qquad\qquad q_1 + q_2 = F(Q_1 - 1) = F(Q_2 - 1)$

In the following it will prove convenient to rewrite (9.7) in the form

$(9.8a)$ $\qquad Q_1 = cq_1 + 1, \qquad q_2 = 0, \qquad$ on face 1

$(9.8b)$ $\qquad q_1 = 0, \qquad Q_2 = cq_2 + 1, \qquad$ on face 2

$(9.8c)$ $\qquad Q_1 = Q_2 = c(q_1 + q_2) + 1, \qquad$ on corner 1-2

where

$(9.8d)$ $\qquad\qquad\qquad c = 1/F$

It must be recalled that Eqs. (9.8) are valid only if the stress point has progressed regularly.

For a kinematic model material * [5.1], it follows from (5.3) and (5.4) applied to this particular example that

$(9.9a)$ $\qquad \dot{q}_1 = \dot{Q}_1/c, \qquad \dot{q}_2 = 0, \qquad$ on face 1

$(9.9b)$ $\qquad \dot{q}_1 = 0, \qquad \dot{q}_2 = \dot{Q}_2/c, \qquad$ on face 2

$(9.9c)$ $\qquad c\dot{q}_1 = \dot{Q}_1, \qquad c\dot{q}_2 = \dot{Q}_2, \qquad$ on corner 1-2

Again assuming a regular progression for the stress point, these equations may be integrated to yield

$(9.10a)$ $\quad Q_1 = cq_1 + 1, \qquad q_2 = 0, \qquad$ on face 1

$(9.10b)$ $\quad q_1 = 0, \qquad Q_2 = cq_2 + 1, \qquad$ on face 2

$(9.10c)$ $\quad Q_1 = cq_1 + 1, \qquad Q_2 = cq_2 + 1, \qquad$ on corner 1-2

Finally, we consider the general linear combination of these two types of hardening [6.3]. In the tension-compression diagram, Fig. 5, let us assume that the actual reversed loading curve is along CJK. Such a material may be considered to consist of a mixture of α isotropic elements and $1 - \alpha$ kinematic model elements,† where α is the

---

* Simple kinematic hardening is considered here, the possible effects of any zero stresses Q_3, ... being neglected.

† This fiction of considering a physical material as composed of a homogeneous mixture of various simple elements has been previously used by White [9.3] and Besseling [6.4].

ratio of DJ to DF. Therefore, multiplying (9.8) by α and (9.10) by $1 - \alpha$ and adding, we obtain the combined law in the form

(9.11a) $Q_1 = cq_1 + 1,$ $q_2 = 0,$ on face 1

(9.11b) $q_1 = 0,$ $Q_2 = cq_2 + 1,$ on face 2

(9.11c) $Q_1 = c(q_1 + \alpha q_2) + 1,$ $Q_2 = c(\alpha q_1 + q_2) + 1,$

on corner 1-2

The general situation involving arbitrary faces and corners can be handled in a similar manner.

CHAPTER 4

Minimum Principles of Plasticity

10. Introduction

In the theory of elasticity, the principles of minimum potential energy and minimum complementary energy have played a very important role. Therefore, it is only natural that attempts have been made to establish similar principles in the theory of plasticity.* The basic concept of these minimum principles is that a class of functions is defined which satisfies some, but not all, of the requirements of a complete solution. It is then shown that a certain functional expression, defined for this class of functions, is a minimum for that function which satisfies the remaining requirements for a complete solution.

The precise form of the minimum principle will depend, of course, on the particular relations between the variables of the problem. However, certain features are common to all of the principles, and other features are clearly analogous. Therefore, although the fact is often obscured in the usual treatment of the minimum principles of elasticity, they are, in fact, the prototypes of most of the principles to be discussed in the following. For this reason, we will begin this chapter on minimum principles with a presentation of the well-known elastic minimum principles from a viewpoint designed to bring out the later analogies.

A typical boundary problem in elasticity consists of being given the displacements on a portion of the boundary and the tractions on the remainder. More generally, a portion of the boundary will be defined as a displacement-type boundary S_D if in each of three independent directions we are either given the component of displacement or else

---

* Although a complete history of the work done in this field would be beyond the scope of this volume, mention may be made of the work of Colonnetti [10.1], Kachanov [24.13], Sadowsky [10.2], Markov [24.18], Hill [10.3, 10.4, 10.5], Phillips [10.6, 10.7], Greenberg [10.8, 10.9], Nadai [10.10], Hodge and Prager [10.11], and Finzi [10.12].

81

the component of traction is required to vanish. Such a boundary is then characterized by the requirement that if $\mathbf{u}_1$ and $\mathbf{u}_2$ are any two vectors which satisfy all boundary conditions on displacement, and $\mathbf{T}$ is any vector which satisfies the boundary conditions on tractions, then

$$(10.1) \qquad \mathbf{u}_1 \cdot \mathbf{T} = \mathbf{u}_2 \cdot \mathbf{T} \text{ on } S_D$$

Similarly, a traction portion of the boundary S_T is defined by the requirement that

$$(10.2) \qquad \mathbf{T}_1 \cdot \mathbf{u} = \mathbf{T}_2 \cdot \mathbf{u} \text{ on } S_T$$

Although these are not the most general conditions possible, it will be assumed in the following that the surface S consists entirely of regions of the type S_D or S_T.

The solution to the elasticity problem is to be expressed in terms of generalized stresses Q_i and generalized strains q_i, which must satisfy certain conditions. The stresses Q_i must satisfy the appropriate equations of internal equilibrium and must be in equilibrium with the applied tractions $\mathbf{T}$ wherever these are prescribed. Any set of functions Q_i^0 which satisfy these conditions will be called "statically admissible." Further, at all boundary points, a vector function $\mathbf{T}^0$ will be defined as the traction across the boundary due to the stress field Q_i^0.

The actual strain field must be derivable from a displacement vector $\mathbf{u}$, so that given q_i, $\mathbf{u}$ can be determined. If $\mathbf{u}^*$ is any displacement vector which satisfies all boundary conditions on $\mathbf{u}$, and the q_i^* are the corresponding generalized strains, then q_i^* is defined as a "geometrically admissible" strain field.

Finally, it will be assumed that the actual stresses and strains are linearly related, so that (1.4) is valid.

Since B_{ij} possesses a unique inverse, statically admissible strains may be defined by

$$(10.3) \qquad q_i^0 = B_{ij}Q_j^0$$

and geometrically admissible stresses by the solution of

$$(10.4) \qquad q_i^* = B_{ij}Q_j^*$$

In general, the q_i^0 cannot be integrated to yield displacements, and the Q_j^* are not in equilibrium.

In terms of the quantities so defined, and in view of (1.1) expressing the relation between Q_i and q_i (where for convenience we have taken $C = 1$), the principle of virtual work may be written in the form †

† For simplicity of exposition we assume there are no body forces.

$$(10.5) \qquad \int_V Q_i{}^0 q_i{}^* \, dV = \int_S \mathbf{T}_i{}^0 \cdot \mathbf{u}^* \, dS$$

where $Q_i{}^0$ is any statically admissible stress field and $q_i{}^*$ is any geometrically admissible displacement field.

The internal potential energy per unit volume associated with any geometrically admissible state is

$$(10.6) \qquad U^* = \int Q_i{}^* \, dq_i{}^* = \tfrac{1}{2} B_{ij} Q_i{}^* Q_j{}^*$$

where the last step follows from (10.4). Therefore, the total potential energy is

$$(10.7) \qquad \Pi^* = \tfrac{1}{2} \int_V B_{ij} Q_i{}^* Q_j{}^* \, dV - \int_{S_T} \mathbf{T} \cdot \mathbf{u}^* \, dS$$

The principle of minimum potential energy then states that among all geometrically admissible displacement fields $\mathbf{u}^*$ the actual one minimizes Π^*. We shall now proceed to prove this theorem.

If the actual potential energy is denoted by Π, then the theorem is equivalent to the statement that

$$\Delta\Pi \equiv \Pi^* - \Pi$$

$$= \tfrac{1}{2} \int_V B_{ij}(Q_i{}^* Q_j{}^* - Q_i Q_j) \, dV - \int_{S_T} \mathbf{T} \cdot (\mathbf{u}^* - \mathbf{u}) \, dS$$

is equal to or greater than zero, with equality only if $\mathbf{u}^* \equiv \mathbf{u}$ to within a rigid body motion. In view of (10.1), (10.4), and (10.5) we may write

$$\int_{S_T} \mathbf{T} \cdot (\mathbf{u}^* - \mathbf{u}) \, dS = \int_S \mathbf{T} \cdot (\mathbf{u}^* - \mathbf{u}) \, dS$$

$$= \int_V Q_i(q_i{}^* - q_i) \, dV$$

$$= \int_V B_{ij} Q_i (Q_j{}^* - Q_j) \, dV$$

Therefore,

$$\Delta\Pi = \tfrac{1}{2} \int_V B_{ij}(Q_i{}^* Q_j{}^* - 2 Q_i Q_j{}^* + Q_i Q_j) \, dV$$

$$(10.8)$$

$$= \tfrac{1}{2} \int_V B_{ij}(Q_i{}^* - Q_i)(Q_j{}^* - Q_j) \, dV$$

where the last step follows from the symmetry of B_{ij}. Finally, since B_{ij} is positive definite, the integrand in (10.8) is nonnegative at each point. Therefore, $\Delta\Pi \geq 0$ with equality if and only if $Q_i^* \equiv Q_i$, which latter imples $q_i^* = q_i$ and hence $\mathbf{u}^* = \mathbf{u}$ to within a rigid body motion. Q.E.D.

Similarly, the internal complementary energy per unit volume associated with any statically admissible state is defined by

$$(10.9) \qquad U_c{}^0 = \int q_i{}^0 \, dQ_i{}^0 = \tfrac{1}{2} B_{ij} Q_i{}^0 Q_j{}^0$$

and the total complementary energy by

$$(10.10) \qquad \Pi_c{}^0 = \tfrac{1}{2} \int_V B_{ij} Q_i{}^0 Q_j{}^0 \, dV - \int_{S_D} \mathbf{T}^0 \cdot \mathbf{u} \, dS$$

The principle of minimum complementary energy then states that among all statically admissible stress states $Q_i{}^0$ the actual one minimizes $\Pi_c{}^0$. The proof is analogous to the previous one and will not be repeated.

Finally, it follows directly from the principle of virtual work that for the actual state, $\Pi + \Pi_c = 0$. Therefore the inequalities implied by the two principles may be combined into the continuing inequality

$$(10.11) \qquad -\Pi_c{}^0 \leq -\Pi_c = \Pi \leq \Pi^*$$

to yield both upper and lower bounds on either of the energies.

All of the conclusions of this section have been formulated in terms of generalized variables. In any specific situation, the choice of variables may implicitly imply further restrictions on the definitions of admissible fields. For example, if the material is assumed to be always incompressible, then only incompressible displacement fields can be considered admissible. Such a condition may also have a bearing on the uniqueness questions discussed in the following sections. Details should be evident in any particular application.

11. Rate principles

For the principles to be proved in the present section, the basic relations concern stress rates and velocities, rather than stresses and displacements. Therefore, we begin by defining a velocity surface S_V by

$$(11.1) \qquad \mathbf{v}_1 \cdot \mathbf{T} = \mathbf{v}_2 \cdot \mathbf{T} \text{ on } S_V$$

and a stress rate surface by

$$(11.2) \qquad \dot{\mathbf{T}}_1 \cdot \mathbf{v} = \dot{\mathbf{T}}_2 \cdot \mathbf{v} \text{ on } S_T$$

The boundary value problems to be solved now presuppose that at a certain time t the displacements and stresses are known throughout the body, and stress rates and velocities are prescribed on S in accordance with (11.1) and (11.2). The problem is to determine the stress rates and velocities throughout the interior V of S.

For a perfectly plastic material, the stress-strain law of the previous section must be replaced by (2.9a): †

$$(11.3a) \qquad \dot{q}_i = B_{ij}\dot{Q}_j + \dot{\lambda}\, \partial f/\partial Q_i$$

In particular, at a plastic point

$$(11.3b) \qquad f = 1, \quad \dot{f} = 0, \quad \text{and } \dot{\lambda} \geq 0$$

and at an elastic point

$$(11.3c) \qquad \dot{\lambda} = 0 \text{ and either } f < 1 \text{ or } \dot{f} < 0$$

A statically admissible field of stress rates $\dot{Q}_i^0$ is now defined as one which is in equilibrium with the applied traction rates $\dot{\mathbf{T}}$, and which does not violate the plasticity conditions. This latter requirement is automatically satisfied if $f < 0$, but imposes the additional qualification

$$(11.4a) \qquad \dot{f}^0 \leq 0 \text{ if } f = 1$$

The related strain rates are then given by

$$(11.4b) \qquad \dot{q}_i^0 = B_{ij}\dot{Q}_j^0 + \dot{\lambda}^0\, \partial f/\partial Q_i$$

where, in view of (11.3b),

$$(11.4c) \qquad \text{if } f = 1 \text{ and } \dot{f}^0 = 0, \text{ then } \dot{\lambda}^0 \geq 0$$

$$(11.4d) \qquad \text{if } f < 1 \ \text{ or } \dot{f}^0 < 0, \text{ then } \dot{\lambda}^0 = 0$$

Similarly, a "kinematically admissible" velocity field is any set of strain rates derived from a velocity vector $\mathbf{v}$ satisfying all velocity boundary conditions. The related stress rates are any solution of

$$(11.5a) \qquad \dot{q}_i^* = B_{ij}\dot{Q}_j^* + \dot{\lambda}^*\, \partial f/\partial Q_i$$

such that

$$(11.5b) \qquad \text{if } f = 1 \text{ and } \dot{f}^* = 0, \text{ then } \dot{\lambda}^* \geq 0$$

$$(11.5c) \qquad \text{if } f < 1 \ \text{ or } \dot{f}^* < 0, \text{ then } \dot{\lambda}^* = 0$$

† For simplicity of exposition, we assume throughout the present section that the yield surface has a uniquely defined normal at each point. Koiter [11.1] has given a general demonstration of the manner in which all such proofs may be extended to include corners in the yield surface.

In both (11.4) and (11.5) it will be noted that quantities such as f and $\partial f/\partial Q_i$ which depend only upon the stress state are, of course, evaluated for the actual given stress state.

Now, if $\dot{q}_i{}^*$ represents any kinematically admissible velocity field, the analogy to the potential energy per unit volume is

$$(11.6) \quad \dot{W}^* = \int \dot{Q}_i{}^* \, d\dot{q}_i{}^* = \int [B_{ij}\dot{Q}_i{}^* \, d\dot{Q}_j{}^* + \dot{Q}_i{}^*(\partial f/\partial Q_i) \, d\dot{\lambda}^*]$$

In the last term of (11.6) it is not necessary to take the differential of $(\partial f/\partial Q_i)$ since this depends only upon the stress and not the stress rates. Further, it follows from (11.5) that

$$(\partial f/\partial Q_i)\dot{Q}_i{}^* \, d\dot{\lambda}^* = \dot{f}^* \, d\dot{\lambda}^* = 0$$

since either $\dot{f}^* = 0$ or $\dot{\lambda}^*$ vanishes identically. Therefore

$$\dot{W}^* = \tfrac{1}{2}B_{ij}\dot{Q}_i{}^*\dot{Q}_j{}^*$$

and the total energy rate is

$$(11.7) \quad \dot{\Lambda}^* = \tfrac{1}{2}\int_V B_{ij}\dot{Q}_i{}^*\dot{Q}_j{}^* \, dV - \int_{S_T} \dot{\mathbf{T}}\cdot\mathbf{v}^* \, dS$$

The first minimum principle for an elastic-perfectly plastic material then states that among all kinematically admissible velocity fields the actual one minimizes $\dot{\Lambda}^*$.

By a series of steps exactly analogous to those used obtaining (10.8) we can show that

$$\Delta\dot{\Lambda} \equiv \dot{\Lambda}^* - \dot{\Lambda}$$

$$= \tfrac{1}{2}\int_V B_{ij}(\dot{Q}_i{}^* - \dot{Q}_i)(\dot{Q}_j{}^* - \dot{Q}_j) \, dV$$

$$(11.8) \qquad\qquad\qquad + \int_V (\partial f/\partial Q_i)\dot{Q}_i(\dot{\lambda} - \dot{\lambda}^*) \, dV$$

Now, the second integrand can be written as $\dot{f}(\dot{\lambda} - \dot{\lambda}^*)$, and it follows from (11.3b) that this vanishes in a region where the material is plastic. On the other hand, if the material is elastic, it follows from (11.3c) that $\dot{\lambda} = 0$ so that the integrand reduces to $-\dot{f}\dot{\lambda}^*$. Finally, if $f < 1$, no finite stress rates can make the neighborhood immediately plastic, so that $\dot{\lambda}^* = 0$, whereas if $f = 1$ we must have $\dot{f} < 0$, $\dot{\lambda}^* \geq 0$, so that the term in question is positive. Since the first integrand is a positive definite form, it follows that $\Delta\dot{\Lambda} \geq 0$. Q.E.D.

In similar fashion the total complementary energy rate is defined by

(11.9) $$\dot{\Lambda}_c{}^0 = \tfrac{1}{2} \int_V B_{ij} \dot{Q}_i{}^0 \dot{Q}_j{}^0 \, dV - \int_{S_V} \dot{\mathbf{T}}^0 \cdot \mathbf{v} \, dS$$

The second minimum principle then states that among all statically admissible stress rate states the actual one minimizes $\dot{\Lambda}_c{}^0$. As above, it is readily shown that

$$\Delta \dot{\Lambda}_c \equiv \dot{\Lambda}_c{}^0 - \dot{\Lambda}_c$$

$$= \tfrac{1}{2} \int_V B_{ij}(\dot{Q}_i{}^0 - \dot{Q}_i)(\dot{Q}_j{}^0 - \dot{Q}_j) \, dV$$

(11.10) $$+ \int_V \dot{\lambda}(\partial f/\partial Q_i)(\dot{Q}_i - \dot{Q}_i{}^0) \, dV$$

Here the second integrand can be written as $\dot{\lambda}(\dot{f} - \dot{f}^0)$. This vanishes in the elastic region, whereas in the plastic region $\dot{\lambda} \geq 0$, $\dot{f} = 0$, and $\dot{f}^0 \leq 0$, so that the integrand is nonnegative. The proof follows as before.

Exactly as in the case of elasticity theory, these two theorems may be combined to yield upper and lower bounds on either $\dot{\Lambda}$ or $\dot{\Lambda}^*$. Thus

(11.11) $$-\dot{\Lambda}_c{}^0 \leq -\dot{\Lambda}_c = \dot{\Lambda} \leq \dot{\Lambda}^*$$

Either of these two principles may also be used to discuss the question of uniqueness of solution.† Suppose, for example, that there exist two complete solutions to the boundary value problem stated, and let $\dot{\Lambda}'$ and $\dot{\Lambda}''$ be the corresponding energy rates. Since any solution minimizes the energy rate, these two quantities must be equal. Therefore, analogously to (11.8) we may write

(11.12) $$\tfrac{1}{2} \int_V B_{ij}(\dot{Q}_i' - \dot{Q}_i'')(\dot{Q}_j' - \dot{Q}_j'') \, dV + \int_V \dot{f}'(\dot{\lambda}' - \dot{\lambda}'') \, dV = 0$$

Since B_{ij} is positive definite, the first integral is nonnegative, so that the second integral must be nonpositive. However, by the argument used following (11.8), the last integrand is positive in a region where the primed state is elastic and the unprimed state is plastic, and is elsewhere zero. Therefore (11.12) rules out the existence of any such region. Since there is no preferred position between the primed and unprimed states, it follows that the elastic-plastic boundary is uniquely

† A quite different approach to uniqueness has been presented by Drucker [11.2]. Other uniqueness questions have been discussed by Hill [11.3].

determined. Finally, since the second integral in (11.12) vanishes, the first must also and hence

$$(11.13) \qquad \dot{Q}_i{}' = \dot{Q}_i{}''$$

Since the elastic stress-strain rate law is one-to-one, the strain rates must also be unique in the elastic region. However, in the plastic region the unknown factor λ occurring in the flow law (11.3a) implies that more than one strain rate state may be associated with the given stress rates. Therefore, the question of uniqueness of the strain rates for a perfectly plastic material must be regarded as still open.

For a strain-hardening material the results are very similar. We consider, for simplicity, the isotropic material. Since instantaneously there is no basic distinction between this and a kinematic model material, the results are quite general. The complete flow law can be stated in the form

$$(11.14a) \qquad \dot{q}_i = B_{ij}\dot{Q}_j + \alpha F(\partial f/\partial Q_i)\dot{f}$$

where

$$(11.14b) \qquad \text{if } f = f_{\max} \geq 1, \dot{f} > 0, \text{ then } \alpha = 1$$

$$(11.14c) \qquad \text{if } f < f_{\max}, \text{ or if } f < 1, \text{ or } \dot{f} \leq 0, \text{ then } \alpha = 0$$

Statically and kinematically admissible fields are defined as before with obvious modifications of (11.4) and (11.5). The expressions for the potential and complementary energy rates turn out to be

$$(11.15) \qquad \dot{\Lambda}^* = \tfrac{1}{2}\int_V (B_{ij}\dot{Q}_i{}^*\dot{Q}_j{}^* + F\alpha^*\dot{f}^{*2})\,dV - \int_{S_T} \dot{\mathbf{T}}\cdot\mathbf{v}^*\,dS$$

$$(11.16) \qquad \dot{\Lambda}_c{}^0 = \tfrac{1}{2}\int_V (B_{ij}\dot{Q}_i{}^0\dot{Q}_j{}^0 + F\alpha^0\dot{f}^{02})\,dV - \int_{S_V} \dot{\mathbf{T}}^0\cdot\mathbf{v}\,dS$$

and the statements of the principles are identical with those following (11.7) and (11.9).

Following the same reasoning as before, we obtain

$$\Delta\dot{\Lambda} = \tfrac{1}{2}\int_V \{B_{ij}(\dot{Q}_i{}^* - \dot{Q}_i)(\dot{Q}_j{}^* - \dot{Q}_j) + F[\alpha^*\dot{f}^*(\dot{f}^* - 2\dot{f}) + \alpha\dot{f}^2]\}\,dV$$

hence it is necessary to show that the term in brackets is nonnegative. We may distinguish four cases, depending upon whether α and α^* are

0 or 1. Denoting the bracket by B^* we obtain

(11.17a)   if $\alpha = \alpha^* = 0$, $B^* = 0$

(11.17b)   if $\alpha = 0$, $\alpha^* = 1$, $B^* = \dot{f}^*(\dot{f}^* - 2\dot{f})$

(11.17c)   if $\alpha = 1$, $\alpha^* = 0$, $B^* = \dot{f}^2$

(11.17d)   if $\alpha = \alpha^* = 1$, $B^* = (\dot{f}^* - \dot{f})^2$

In (11.17b), $\alpha^* = 1$ implies that $f = f_{\max}$ since otherwise both states would be elastic. It then follows from (11.14) that $\dot{f}^* \geq 0$, $\dot{f} \leq 0$, so that $B^* \geq 0$. Obviously the same result is true in the other three situations and the proof of the theorem continues as in the other cases.

In similar fashion, it can be shown that the proof of the second principle depends upon showing that

(11.18) $$B^0 = \alpha^0 \dot{f}^{02} + \alpha(\dot{f}^2 - 2\dot{f}\dot{f}^0)$$

is positive, and a consideration of the same four cases shows that this is indeed so.

Here again, it follows from the flow law (11.14a) that for the actual state $\dot{\Lambda} + \dot{\Lambda}_c = 0$, so the continued inequality (11.11) is valid in this case also. Regarding uniqueness, the same argument used previously shows that the stress rates are unique, and in this case (11.14a) also predicts unique strain rates.

12. Finite principles

Corresponding minimum principles exist for the integrated laws discussed in Chapter 3, although in much more restrictive form. These restrictions require that the stress point in a statically or geometrically admissible stress field have a history of regular progression. It must be noted that this restriction applies not only to the assumed stress or displacement field which is being tested but also to the actual one. Since the actual behavior is unknown, this limits the confidence with which the principles can be used.

With this additional requirement, the statement of the principles is identical with that given in Section 10, except that suitable definitions must be given to Π^* and Π_c^0. Considering first a perfectly plastic material, we see from (8.8) that

$$U^* = \int Q_i^* \, dq_i^* = \int Q_i^* (B_{ij} Q_j^* + \Sigma \, d\lambda_\beta^* a_{\beta i})$$

(12.1)

$$= \tfrac{1}{2} B_{ij} Q_i^* Q_j^* + \Sigma \lambda_\beta^*$$

Therefore, the total potential is

$$\Pi^* = \int_V (\tfrac{1}{2}B_{ij}Q_i{}^*Q_j{}^* + \Sigma\lambda_\beta{}^*)\,dV - \int_{S_T} \mathbf{T}\cdot\mathbf{u}^*\,dS \tag{12.2}$$

Similarly, for a statically admissible field

$$U_c{}^0 = \int q_i{}^0\,dQ_i{}^0 = \int (B_{ij}Q_j{}^0\,dQ_i{}^0 + \Sigma\lambda_\beta a_{\beta i}\,dQ_i{}^0) \tag{12.3}$$

Now, if face β is plastic it follows that $a_{\beta i}\,dQ_i{}^0 = d(1) = 0$, and otherwise $\lambda_\beta = 0$, so that $U_c{}^0$ is still given by (10.9), and

$$\Pi_c{}^0 = \tfrac{1}{2}\int_V B_{ij}Q_i{}^0Q_j{}^0\,dV - \int_{S_D} \mathbf{T}^0\cdot\mathbf{u}\,dS \tag{12.4}$$

In (12.2) the quantities $Q_i{}^*$ and $\lambda_\beta{}^*$ are defined by equations analogous to (8.12) and (8.13), whereas the quantities in (12.4) are obtained directly from the given stress field.

The proof of the theorems proceeds in the usual manner. Thus, it is required to show that

$$\Delta\Pi = \int_V [\tfrac{1}{2}B_{ij}(Q_i{}^*Q_j{}^* - Q_iQ_j) + \Sigma(\lambda_\beta{}^* - \lambda_\beta)]\,dV - \int_{S_T} \mathbf{T}\cdot(\mathbf{u}^* - \mathbf{u})\,dS$$

is nonnegative. Application of the principle of virtual work and the stress-strain law (8.8) reduces this to

$$\Delta\Pi = \int_V [\tfrac{1}{2}B_{ij}(Q_i{}^* - Q_i)(Q_j{}^* - Q_j) + \Sigma(\lambda_\beta{}^* - \lambda_\beta)$$
$$\qquad\qquad\qquad - \Sigma\lambda_\beta{}^*a_{\beta i}Q_i + \Sigma\lambda_\beta a_{\beta i}Q_i\,dV \tag{12.5}$$

Now the summations of $\lambda_\beta{}^*$ are to be taken over those faces which are plastic according to the assumed geometrically admissible state, and the summations of λ_β are over those faces which are actually plastic. In the latter case, $a_{\beta i}Q_i = 1$ so that the λ_β terms cancel out and (12.5) can be written

$$\Delta\Pi = \int_V [\tfrac{1}{2}B_{ij}(Q_i{}^* - Q_i)(Q_j{}^* - Q_j) + \Sigma\lambda_\beta{}^*(1 - a_{\beta i}Q_i)]\,dV$$

which is obviously nonnegative since the actual stress point does not violate any of the yield conditions.

Similarly, we may show that the proof of the principle of minimum complementary energy reduces to demonstrating that

(12.6) $\Delta\Pi_c = \int \left[\frac{1}{2}B_{ij}(Q_i{}^0 - Q_i)(Q_j{}^0 - Q_j) + \Sigma\lambda_\beta(1 - a_{\beta i}Q_i{}^0)\right] dV$

is nonnegative. The same argument as above leads to the desired conclusion.

The two principles can be combined to prove that (10.11) is valid in this case also. Finally, an argument similar to that following (11.11) shows that the stresses are unique but that a certain ambiguity may be present in the strains.

The same two principles are valid for the strain-hardening material described in Section 9. In the case of isotropic hardening [6.1, 6.2], the energy expressions are

(12.7) $\Pi^* = \frac{1}{2}\int [B_{ij}Q_i{}^*Q_j{}^* + F(f^{*2} - 1)]\, dV - \int_{S_T} \mathbf{T}\cdot\mathbf{u}^*\, dS$

and

(12.8) $\Pi_c{}^0 = \frac{1}{2}\int [B_{ij}Q_i{}^0Q_j{}^0 + F(f^0 - 1)^2]\, dV - \int_{S_V} \mathbf{T}^0\cdot\mathbf{u}\, dS$

The proofs are analogous to those presented previously and will not be repeated here. The continued inequality (10.11) is again valid, and the existence of a minimum provides uniqueness of both stresses and strains. Similar results are available for the kinematic model and combined hardening materials.

13. Limit analysis †

The principles of limit analysis represent a different type of minimum principle. Whereas the principles of minimum potential and complementary energy regard the tractions as fixed and minimize the energy, we shall now be concerned with certain multipliers of the tractions. Thus, if a perfectly plastic solid is to be loaded by a set of tractions $\mathbf{T}$ we consider the tractions $\lambda\mathbf{T}$ as λ is slowly increased from zero. In particular, we assume that the entire surface is of the type S_T so that either the component of $\mathbf{T}$ is prescribed or the corresponding component of $\mathbf{v}$ vanishes.

Now, if we assume that the deformations are sufficiently small so that the loads may continue to be applied to the undeformed body, we may, in general, distinguish distinct ranges of value for λ. For λ suf-

---

† The principles of limit analysis were first formulated by Hill [13.1, 13.2] for a rigid-plastic material, and by Drucker, Greenberg, and Prager [13.3, 13.4, 13.5] for an elastic-plastic material (see also [1.1]). The relation between these theorems and complete solutions of perfectly plastic problems has been discussed by Bishop [13.6] and by Lee and others [13.7, 13.8].

ficiently small, say $0 < \lambda < \lambda_E$, the body will be entirely elastic. As λ increases above λ_E part of the body will become plastic, but sufficient elastic material will remain to support the additional load. However, for a certain critical value $\lambda = S$, the elastic material will no longer be sufficient to support any further load, with the result that the strains can increase indefinitely with no increase in load. The value S is known as the safety factor under the given loads $\mathbf{T}$.

For the principles of limit analysis a statically admissible stress field is defined as one which is in equilibrium with the surface tractions $m_S\mathbf{T}$ and nowhere violates the yield condition. Further, the factor m_S will be defined as a statically admissible multiplier provided there exists at least one statically admissible stress field associated with it. The first principle of limit analysis then states that the safety factor is the largest statically admissible multiplier.

Before proving this principle we shall show that under the loads $S\mathbf{T}$ the stress rates $\dot{Q}_i$ vanish throughout the body. We observe first that under the loads $S\mathbf{T}$

$$(13.1) \qquad \int_V \dot{Q}_i\dot{q}_i \, dV = \int_S S\dot{\mathbf{T}}\cdot\mathbf{v} \, dS = 0$$

where the conclusion follows from the definition of S as the load multiplier under which deformations can continue with no change in traction. However, it follows from the flow law (2.9b) that

$$\int_V \dot{Q}_i\dot{q}_i \, dV = \int_V \dot{Q}_i(B_{ij}\dot{Q}_j + \Sigma_\alpha\lambda_\alpha \, \partial f_\alpha/\partial Q_i) \, dV$$

$$(13.2) \qquad\qquad = \int_V (B_{ij}\dot{Q}_i\dot{Q}_j + \Sigma_\alpha\dot{\lambda}_\alpha\dot{f}_\alpha) \, dV$$

Corresponding to each plastic face, either $\dot{\lambda}_\alpha = 0$ if that face does not enter into the flow mechanism, or else the corresponding $\dot{f}_\alpha = 0$. Therefore, combining (13.1) and (13.2) we obtain

$$(13.3) \qquad \int_V B_{ij}\dot{Q}_i\dot{Q}_j \, dV = 0$$

Now, since B_{ij} is positive definite this implies that $\dot{Q}_i = 0$ at each point of V. Q.E.D. Finally, the substitution of this result into (2.9b) shows that under the loads $S\mathbf{T}$ the flow law may be written

$$(13.4) \qquad \dot{q}_i = \Sigma\lambda_\alpha \, \partial f_\alpha/\partial Q_i$$

Returning now to the first principle, we consider

$$\int_V (Q_i - Q_i^0)\dot{q}_i \, dV$$

where Q_i^0 is a statically admissible stress field. Now, according to the reasoning of Section 2 [see (2.8)], this integrand is nonnegative at each point. Therefore,

$$\int_V (Q_i - Q_i^0)\dot{q}_i \, dV = \int_S (S\mathbf{T} - m_S\mathbf{T}) \cdot \mathbf{v} \, dS$$

(13.5)
$$= (S - m_S) \int_S \mathbf{T} \cdot \mathbf{v} \, dS \geq 0$$

Since $\int_S \mathbf{T} \cdot \mathbf{v} \, dS$ represents the plastic work and is positive, it follows that $S \geq m_S$. Therefore, since S is by definition a statically admissible multiplier, it is the largest such multiplier. Q.E.D.

The second principle is stated in terms of a kinematically admissible velocity field. Let $\mathbf{v}^*$ be any field which satisfies any zero velocity boundary conditions and the requirement that $\int_S \mathbf{T} \cdot \mathbf{v}^* > 0$. Now, assuming that $\mathbf{v}^*$ is entirely plastic, a strain rate field $\dot{q}_i^*$ and stress field Q_i^* can be associated therewith as indicated in Fig. 12. In general,

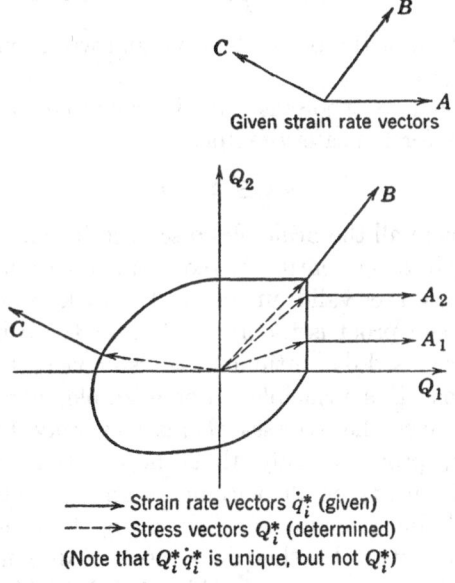

Fig. 12. Determination of kinematically admissible stress fields.

this stress field will not be in equilibrium, and it may not be unique. However, we may define a unique kinematically admissible multiplier m_k by

(13.6)
$$m_k = \frac{\displaystyle\int_V Q_i{}^* \dot{q}_i{}^* \, dV}{\displaystyle\int_S \mathbf{T} \cdot \mathbf{v}^* \, dS}$$

since $Q_i{}^* \dot{q}_i{}^*$ is unique, even if $Q_i{}^*$ is not. The second principle of limit analysis then states that the safety factor is the smallest kinematically admissible multiplier.

For proof, we consider

$$\int_V (Q_i{}^* - Q_i) \dot{q}_i{}^* \, dV$$

Again, by the arguments of Section 2 the integrand is everywhere non-negative so that

$$\int_V (Q_i{}^* - Q_i) \dot{q}_i{}^* \, dV = \int_S (m_k \mathbf{T} - S\mathbf{T}) \cdot \mathbf{v}^* \, dS$$

(13.7)
$$= (m_k - S) \int_S \mathbf{T} \cdot \mathbf{v}^* \geq 0$$

so that $m_k \geq S$. Q.E.D. Observe that the transformation of $\displaystyle\int_V Q_i{}^* \dot{q}_i{}^* \, dV$ is not an application of the principle of virtual work, but a consequence of the definition (13.6).

Obviously these two theorems may be combined to provide upper and lower bounds for the safety factor:

(13.8)
$$m_S \leq S \leq m_k$$

In this proof, as in all the principles discussed in this chapter, we have tacitly assumed whatever continuity requirements are necessary. Actually, the theorems are valid under fairly weak requirements. The crucial point in each proof is the application of the principle of virtual work, and any proposed discontinuous stress or velocity field need only be checked against this principle. For example, not only the stress derivatives but even the stresses themselves may be discontinuous across a surface, provided only that the tractions are continuous. Continuity requirements on displacements and velocities can also be relaxed, provided that account is taken of any finite energy produced at such a discontinuity. Further discussions of discontinuities in perfectly plastic solids may be found in [13.9, 13.10, 13.11, 13.12].

CHAPTER 5

BENDING OF A CIRCULAR PLATE

14. Rigid-perfectly plastic material

The first four chapters of this survey have been devoted to the current theories of plasticity, rather than to a list of solutions of problems. Although it must be admitted that there are probably 20 problem solutions published for every paper containing a significant advance in theory, this primary emphasis on theory appears to be justified. For without the theory solutions would be impossible, whereas with the theory solutions are primarily a matter of time and interest. However, we do wish to consider a few particular problems both for the purpose of illustrating the theory and also to give an indication of the present state of development in the field. In this latter connection, no claim of completeness is made—the examples chosen are merely illustrative.

The problem of a circular plate under a radially symmetric load is in many ways an ideal one for our first purpose, since it contains most of the essential ingredients of a more complex problem and yet is mathematically simple. For definiteness we shall consider a simply supported circular plate subjected to a uniform normal pressure p. This and other plate problems were first investigated by Hopkins and Prager [14.1] and later by Hopkins and others [14.2, 14.3]. Plate problems have also been investigated by several Russian authors as listed in the bibliography on pages 138 and 143.

The generalized stresses and strains for this problem were previously defined by (1.2). The stress resultants must satisfy the equilibrium equation

$$(14.1) \qquad (\xi Q_1)' - Q_2 = -3P\xi^2$$

where ξ and P are the dimensionless co-ordinate and pressure defined by

$$(14.2) \qquad \xi = r/a, \qquad P = pa^2/6Yh^2$$

Primes indicate differentiation with respect to ξ. Further, if the dimensionless displacement is given by

(14.3a) $$W = w/a$$

then it follows from (1.2) that the strain variables are

(14.3b) $$q_1 = a\kappa_r = -W'', \qquad q_2 = a\kappa_\theta = -W'/\xi$$

If the material of the plate satisfies Tresca's condition of maximum shearing stress, then it may be shown [14.1] that the moments M_r and M_θ satisfy a geometrically similar restriction. In terms of Q_1 and Q_2 then, the yield condition is

(14.4) $$\max\,[\,|Q_1|,\ |Q_2|,\ |Q_1 - Q_2|\,] \leq 1$$

The corresponding yield frame is shown in Fig. 13.

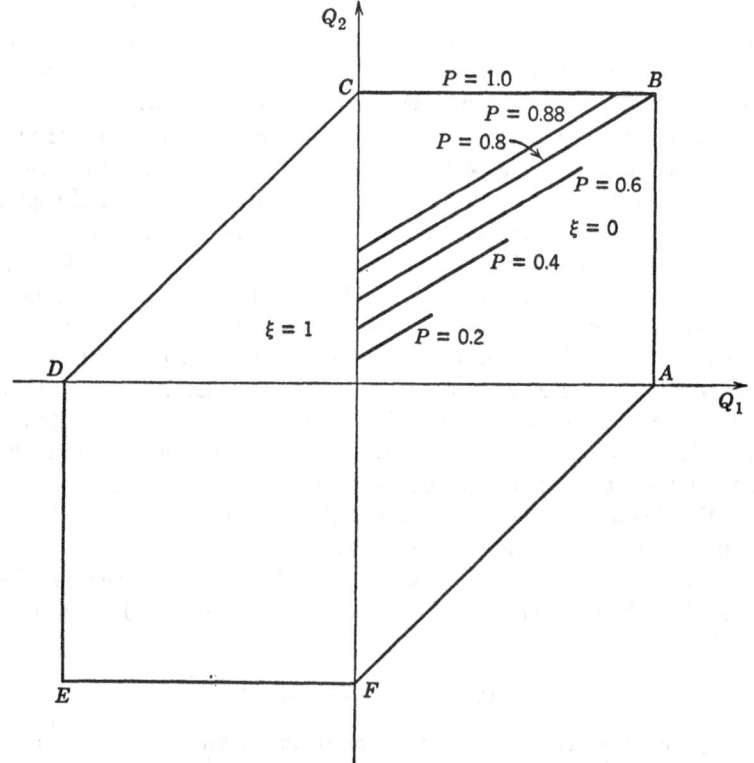

Fig. 13. Yield frame and stress profiles for circular plate.

The general method of solution for a plastic rigid material is to assume a "stress profile" for the body, i.e., to assume the locus of stress points in relation to the yield frame. In particular, it is evident that for P less than some critical value P_0 there will be no deformations and the plate will be rigid, but at $P = P_0$ motion may take place. Obviously, then, the only solution of interest for the rigid-perfectly plastic material is when $P = P_0$.

In order to show how an incorrect hypothesis with regard to the stress profile quickly manifests itself, let us assume first that some finite part of the stress profile is on side AB. It then follows from the flow law that

$$\dot{q}_1 = -\dot{W}'' > 0, \qquad \dot{q}_2 = -\dot{W}'/\xi = 0$$

a result which is patently absurd. Alternatively, suppose that a finite portion of the stress profile is at point B. Then, $Q_1 = Q_2 = 1$ and the equilibrium equation (14.1) reduces to

$$P = 0$$

which is also unacceptable for the present problem.

The correct hypothesis for the stress profile must, of course, satisfy the boundary conditions. At the simply supported edge these are

(14.5a, b) $\qquad\qquad Q_1(1) = 0, \qquad W(1) = 0$

whereas at the center of the plate isotropy demands that

(14.5c) $\qquad\qquad Q_1(0) = Q_2(0)$

If we now assume that the entire plate is plastic under the load P_0, it appears reasonable to take the stress profile to be the line CB, with B corresponding to $\xi = 0$ and C to $\xi = 1$. Therefore, $Q_2 = 1$ throughout the plate, and (14.1), together with the boundary condition (14.5c), yields

$$Q_1 = 1 - P\xi^2$$

Condition (14.5a) then shows that P must equal 1 so that the proposed stress solution is

(14.6) $\qquad\qquad Q_1 = 1 - \xi^2, \qquad Q_2 = 1, \qquad P_0 = 1$

Obviously this solution does not violate the yield condition.

The velocity field associated with (14.6) is obtained from the condition that

(14.7) $\qquad\qquad \dot{q}_1 = -\dot{W}'' = 0, \qquad \dot{q}_2 = -\dot{W}'/\xi > 0$

Therefore $\dot{W}$ is a linear function, and in view of (14.5b) the velocity field is

$$(14.8) \qquad \dot{W} = \dot{W}_0(1 - \xi)$$

Here $\dot{W}_0$ is an undetermined positive constant.

It will be observed that the slope is discontinuous at $\xi = 0$. This, in fact, is a price one must pay for considering such an idealized material. For a real material, such a discontinuity must correspond to a small region of rapidly varying slope, hence of large curvature. In the limit, the curvature must become infinite, in violation of the first Eq. (14.7). However, it will be recalled that the origin is at point B, where (14.7) must be replaced by the conditions

$$\dot{q}_1 = -\dot{W}'' \geq 0, \qquad \dot{q}_2 = -\dot{W}'/\xi \geq 0$$

These conditions are, in fact, satisfied by a negatively infinite $\dot{W}''$. Therefore, the solution given by (14.6) and (14.8) satisfies all the conditions of the problem and is the desired complete solution.

15. Elastic-perfectly plastic material

If the value of P is less than P_0, then the rigid-plastic plate discussed in the previous section will not undergo any deformations, and it will be impossible to analyze the stresses. Therefore, if we wish to obtain information in this case we must consider the elastic strains. This has previously been done by Naghdi [15.1], Hopkins [15.2], Haythornthwaite [15.3], and Tekinalp [15.4].

We define dimensionless variables as before, and assume that the stress-strain diagram *in terms of these variables* is elastic-perfectly plastic. It should be pointed out that this is equivalent to assuming an elastic-perfectly plastic material only in the particular case of a plate of an idealized sandwich construction. However, even for the plate with uniform cross section it may give a reasonable approximation to the actual solution.

Now, for P sufficiently small, the plate will be everywhere elastic. In this case the solution is governed throughout by (1.6), (14.1), and (14.3). If we first solve (1.6) and (14.3) for the stresses, we obtain

$$(15.1) \qquad \begin{aligned} Q_1 &= -(1/b)(W'' + \nu W'/\xi) \\ Q_2 &= -(1/b)(\nu W'' + W'/\xi) \end{aligned}$$

where b is defined by

$$(15.2) \qquad b = (3a/2h)(Y/E)(1 - \nu^2)$$

Substitution of these values into (14.1) then yields a third-order equation in W, the general solution of which is readily verified to be

(15.3) $$W = b(A + B \log \xi + C\xi^2 + \tfrac{3}{32} P\xi^4)$$

Finally, the constants are determined from the boundary conditions (14.5). The complete elastic solution may then be written

$$W = \frac{3Pb}{32}\left(\frac{5+\nu}{1+\nu} - 2\frac{3+\nu}{1+\nu}\xi^2 + \xi^4\right)$$

(15.4) $$Q_1 = \frac{3(3+\nu)}{8} P(1 - \xi^2)$$

$$Q_2 = \frac{3(3+\nu)}{8} P\left(1 - \frac{1+3\nu}{3+\nu}\xi^2\right)$$

Equations (15.4) will be valid for all values of P such that the point with co-ordinates Q_1, Q_2 lies within the yield frame of Fig. 13. This discussion may conveniently be carried on in terms of the stress profile, which is obtained by plotting the curve $Q_1 = Q_1(\xi)$, $Q_2 = Q_2(\xi)$, $0 \le \xi \le 1$, as given by (15.4). Typical stress profiles are shown in Fig. 13, where, for definiteness, ν is taken to be $\tfrac{1}{3}$. Obviously, the critical point in the plate is at the center, so that the plate will remain elastic as long as $Q_1 = Q_2 < 1$ at $\xi = 0$. Thus the maximum elastic load is

(15.5) $$P^* = 8/[3(3 + \nu)]$$

For $P^* < P < P_0$, the plate will evidently be plastic in a central portion $0 < \xi < y$ and elastic for $y < \xi < 1$, where y is to be determined. Equation (15.3) is still valid for the elastic region, but the boundary conditions at $\xi = 0$ will no longer apply. However, Eqs. (14.5a, b) at $\xi = 1$ are still applicable, so that we may write the complete elastic solution in the form

(15.6a)

$$W = b\left[B \log \xi + \left(\frac{1}{2}B\frac{1-\nu}{1+\nu} - \frac{3}{16}\frac{3+\nu}{1+\nu}P\right)(\xi^2 - 1) + \tfrac{3}{32}P(\xi^4 - 1)\right]$$

(15.6b) $Q_1 = (1-\nu)B/\xi^2 - B(1-\nu) + \tfrac{3}{8}(3+\nu)P - \tfrac{3}{8}(3+\nu)P\xi^2$

(15.6c) $Q_2 = -(1-\nu)B/\xi^2 - B(1-\nu) + \tfrac{3}{8}(3+\nu)P - \tfrac{3}{8}(1+3\nu)P\xi^2$

By the same arguments used in the preceding section it can be shown that the only reasonable hypothesis for the plastic stress profile is that it lie along BC in Fig. 13. Therefore, Q_2 has the constant value of 1 and the equilibrium equation (14.1) can be solved to yield the complete stress solution

(15.7a) $\qquad Q_1 = 1 - P\xi^2, \qquad Q_2 = 1, \qquad 0 < \xi < y$

Here the integration constant was determined so as to satisfy the isotropy condition (14.5c) at the plate center.

In view of (1.7) and (8.4), the stress-strain law in the plastic region is

$$q_1 = \frac{b}{1 - \nu^2}(Q_1 - \nu Q_2), \qquad q_2 = \frac{b}{1 - \nu^2}(Q_2 - \nu Q_1) + \lambda$$

The second of these equations serves only to determine λ, and need not further concern us except to verify later that λ is positive. Substitution of (15.7a) and the first Eq. (14.3b) into the expression for q_1 yields an explicit formula for W''. Therefore, a double integration gives the displacement in the plastic region as

(15.7b) $\qquad W = \frac{b}{1 - \nu^2}[F + D\xi - \frac{1}{2}(1 - \nu)\xi^2 + \frac{1}{12}P\xi^4]$

It now remains to determine the values of B, D, F, and y in (15.6) and (15.7). To do this, we have the continuity conditions at $\xi = y$, namely that displacement, slope, and both moments must be continuous. The dependence of the four resulting equations on y is obviously nonlinear, so we adopt the artifice of regarding y as independent and P as dependent. In other words, we determine that value of P for which the elastic-plastic boundary will have a given position. For the particular case $\nu = \frac{1}{3}$ the resulting values of the constants are

(15.8a) $\qquad P = 4(5 - 2y^2 + y^4)^{-1}$

(15.8b) $\qquad B = -\frac{3}{2}y^4(5 - 2y^2 + y^4)^{-1}$

(15.8c) $\qquad D = -\frac{8}{3}y^3(5 - 2y^2 + y^4)^{-1}$

(15.8d) $\qquad F = \frac{1}{3}(4 + 7y^4 - 4y^4 \log y)(5 - 2y^2 + y^4)^{-1}$

Equations (15.6), (15.7), and (15.8) now give the complete elastic-plastic solution as a function of y, and hence, in view of (15.8a), as a function of the load.

For $y = 0$, (15.8a) gives $P = 0.8$, in agreement with (15.5); for $y = 1$ we obtain $P = 1$. According to (14.6) this is the load at which the entire plate becomes plastic, so that we have found the complete elastic-plastic solution.

Salient features of the solution are shown in Figs. 13 through 15. In particular, Fig. 13 shows the stress profile at the intermediate value

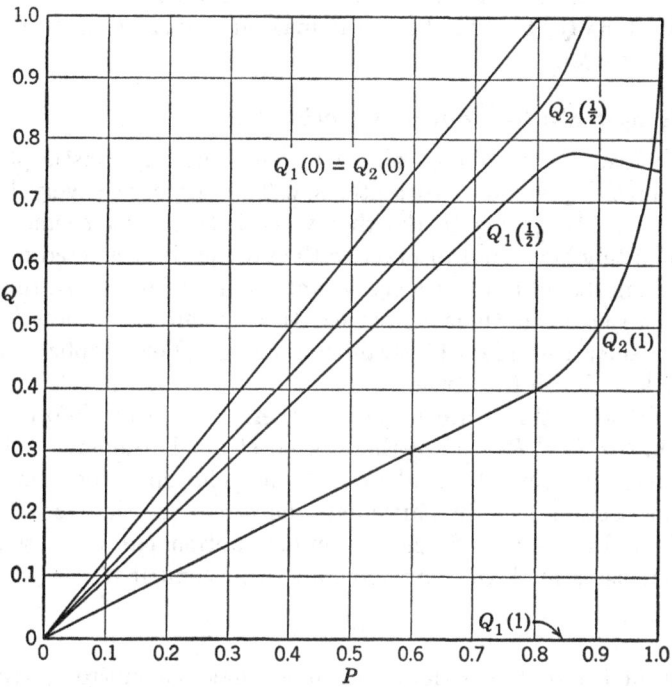

Fig. 14. Bending moments in elastic-plastic plate as a function of load.

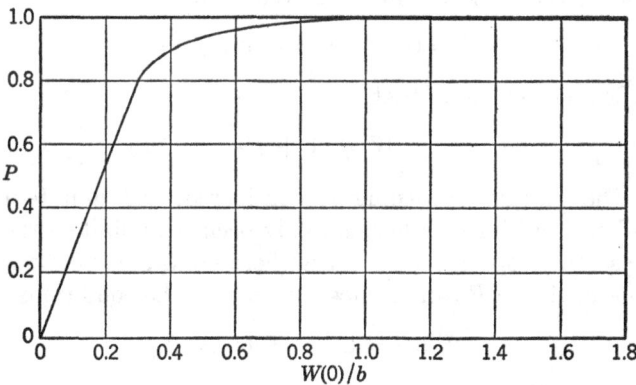

Fig. 15. Relation between load and maximum displacement of circular plate.

$P = 0.88$. Figure 14 shows some of the bending moments as functions of load. Finally, Fig. 15 shows the maximum displacement $W(0)$ as a function of load.

16. Rigid-strain hardening material *

If the load P is increased above P_0, the perfectly plastic plates discussed in the previous two sections will undergo relatively large deformations. According to the theory used, these displacements could be indefinitely large, but in practice the previously neglected membrane forces would begin to exert a significant influence [16.3]. In the present section we shall continue to neglect these membrane forces, but we shall consider the effect of strain hardening. For simplicty we shall also neglect the elastic strains.

Since elastic strains are neglected, there can be no deformations of the shell for $P < P_0$, hence there is no strain hardening. Therefore, up to and including the load $P_0 = 1$, the solution is the same as that given in Section 14. In particular, the stress solution for $P = P_0$ is still given by (14.6). Since continuing motion can only take place with increasing load, the corresponding displacement is

$$(16.1) \qquad\qquad W = 0$$

As a first hypothesis, let us assume that the entire stress profile continues to lie on the side BC of the yield frame (Fig. 13). This is the same situation discussed in Section 9, so the stress-strain law is given by (9.11b). Using (14.3b) to express the strains in terms of W, we may write the stress-strain law in the form

$$(16.2) \qquad\qquad W'' = 0, \qquad Q_2 = -cW'/\xi + 1$$

The first Eq. (16.2) shows that

$$(16.3) \qquad\qquad W = A + B\xi$$

Now, when strain hardening is considered, the infinite curvature associated with a hinge circle could only occur at infinite stress, so that in this case the slope must be everywhere continuous. Since no non-zero values of A and B can be chosen to satisfy the conditions

$$(16.4) \qquad\qquad W'(0) = 0, \qquad W(1) = 0$$

our hypothesis for the solution is incorrect.

---

* For the particular case of simple kinematic model hardening, this problem has been treated by Prager [16.1] and Boyce [16.2]. The present more general treatment is taken from [6.3].

In searching for a second hypothesis, we note that we can still apply the arguments advanced in Section 14 against the use of side AB but not those against the use of corner B. Further, we intuitively expect Q_1 to vanish only at $\xi = 1$ and to be positive elsewhere. Therefore, the reasonable hypothesis appears to be that the stress profile is on AB for $y < \xi < 1$, whereas for $0 < \xi < y$ it is in corner B.

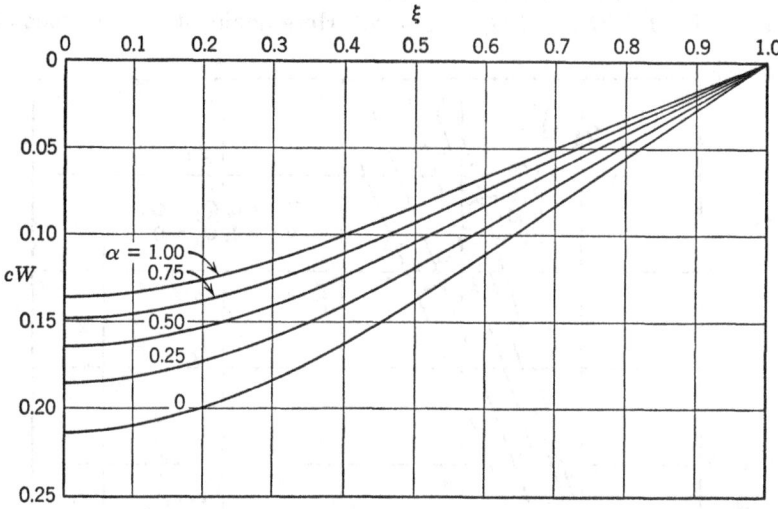

Fig. 16.   Deflected shape of plate for $P = 1.5$.

For $y < \xi < 1$, (16.2) and (16.3) are still valid, and the second boundary condition (16.4) shows that $B = -A$. Substitution of these equations into the equilibrium equation (14.1) furnishes an equation for Q_1. This is easily solved, and the constant of integration is determined from the condition $Q_1(1) = 0$. The complete solution for this region may then be written

$$y < \xi < 1: \quad W = A(1 - \xi)$$

(16.5)
$$Q_1 = 1 - P\xi^2 + (P - 1)/\xi + (cA/\xi) \log \xi$$

$$Q_2 = 1 + cA/\xi$$

For $0 < \xi < y$ the appropriate stress-strain law is given by (9.11c). Therefore, in view of (14.3b),

(16.6)
$$Q_1 = -c(W'' + \alpha W'/\xi) + 1$$

$$Q_2 = -c(\alpha W'' + W'/\xi) + 1$$

The substitution of (16.6) into the equilibrium equation (14.1) yields a third-order equation for W whose solution is

$$(16.7) \qquad cW = C + D\xi^2 + \tfrac{3}{32}P\xi^4$$

Here one constant was determined so as to satisfy the first condition (16.4). Substitution into (16.6) then furnishes the stresses.

The constants A, C, D, and y are to be determined from the continuity of W, W', Q_1, and Q_2 at $\xi = y$. Here again, it is convenient to

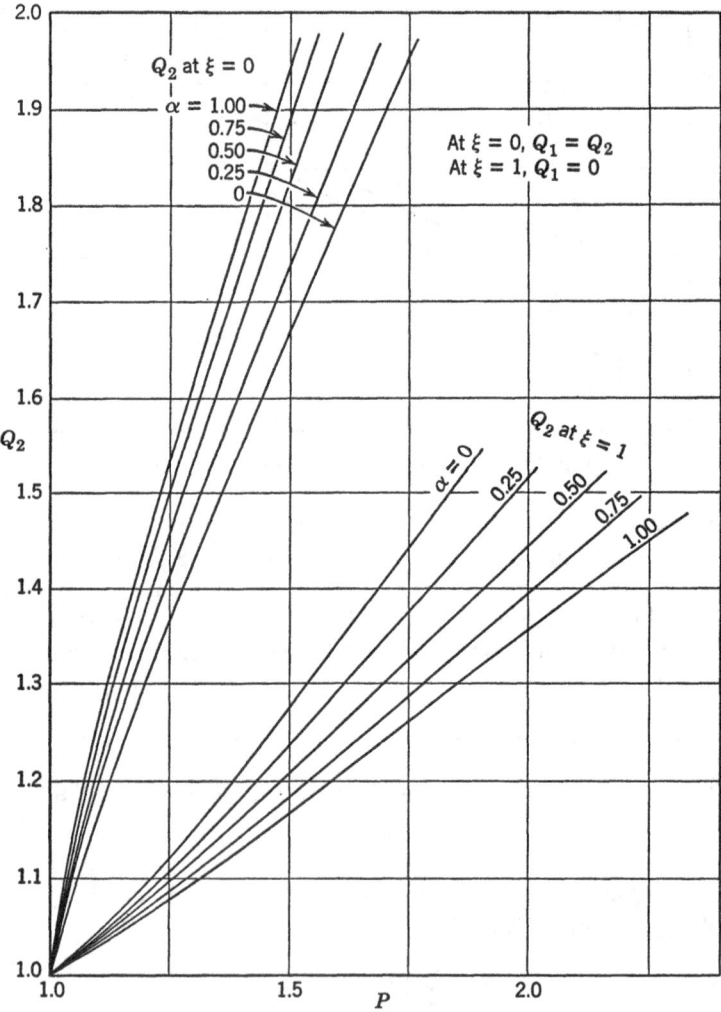

Fig. 17. Bending moments at center and edge as a function of load.

regard y as independent and P as dependent. Using these conditions the complete solution may be written

$$0 \leq \xi \leq y: \quad cW = (3P/32)(8y^3 - 3y^4 - 6y^2\xi^2 + \xi^4)$$

$$Q_1 = 1 + (3P/8)[3(y^2 - \xi^2) + \alpha(3y^2 - \xi^2)]$$

$$Q_2 = 1 + (3P/8)[3\alpha(y^2 - \xi^2) + (3y^2 - \xi^2)]$$

(16.8)

$$y \leq \xi \leq 1: \quad cW = (3P/4)y^3(1 - \xi)$$

$$Q_1 = 1 - P\xi^2 + (P - 1)/\xi + (3P/4)(y^2/\xi) \log \xi$$

$$Q_2 = 1 + (3Py^3/4\xi)$$

$$P = 4[3y^3 \log y + 4 - (4 + 3\alpha)y^3]^{-1}$$

Salient features of the solution are shown in Figs. 16 through 18, which are taken from [6.3].

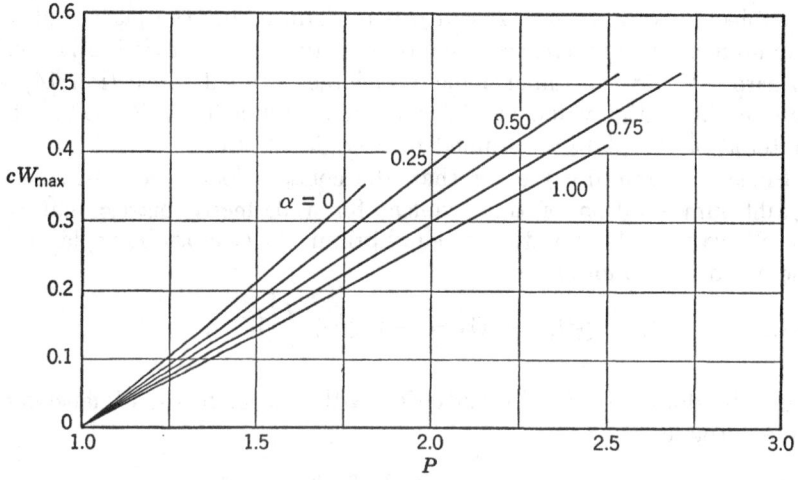

Fig. 18. Maximum plate displacement as a function of load.

The preceding discussion has been based upon a simple kinematic hardening model in which only the moments M_r and M_ϕ are considered. For complete kinematic hardening, it is obviously necessary to consider the direct stress N_r and N_ϕ, but for a symmetric plate cross section it is easily shown that these do not affect the results. There is, however, a more subtle suppressed stress variable Q_3 which is defined by

(16.9)
$$Q_3 = (1/Yh^2)\int_{-h}^{h} z\sigma_z \, dz$$

Since σ_z vanishes, Q_3 is, of course, equal to zero. However, if the Q_3 direction is included in defining the yield frame, it is seen that Fig. 13 is a nonorthogonal section of an equilateral hexagonal cylinder whose axis is the line $Q_1 = Q_2 = Q_3$. The resulting analysis has been treated in [5.2].

17. Dynamic loading

Although dynamic loading generally is beyond the scope of this survey, there is one particular type of dynamic problem which has attracted considerable interest of late. In general, if a structure made of a perfectly plastic material is subjected to a load greater than its yield load, that structure will suffer such large deformations as to become useless. However, if the load is applied for a sufficiently short time, the inertia of the deforming structure may be sufficient to keep the deformations within reasonable limits.

As a typical example,* let us assume that a simply supported plate is subjected to a pressure P which is uniform across the plate, but is a monotonically nonincreasing function of time which eventually tends to zero. Using the dimensionless variables defined in Section 14, we denote the initial value of P by B, and assume that $B > P_0$. The material of the plate is assumed to be rigid plastic.

Since P is initially greater than the collapse load, there will be no equilibrium solution of the problem for a perfectly plastic material. It follows that the equation of equilibrium (14.1) must be replaced by the equation of motion

$$(17.1) \qquad (\xi Q_1)' - Q_2 = -3P\xi^2 + \int_0^\xi \ddot{W}\xi \, d\xi$$

Here the dots indicate differentiation with respect to the dimensionless time variable

$$(17.2) \qquad \tau = (Yh^2/a^3\sigma)^{\frac{1}{2}}t$$

σ being the surface density of the material.

As our initial hypothesis we shall assume that the entire plate is in regime BC during the entire deformation. Since the effect of inertia terms on the stress-strain law has not been previously accounted for, we shall start with the flow law (2.9a). On side BC this reduces to

$$(17.3a, b) \qquad \dot{W}'' = 0, \qquad \dot{W}' \leq 0$$

---

* Dynamic loading of plates was first considered by Hopkins and Prager [17.1]. Other examples have been considered by Hopkins and Wang [17.2, 17.3].

Therefore, since $W(1) = 0$, we may write

(17.4) $W = \phi(\tau)(1 - \xi)$

where ϕ is a positive function of time to be determined.

On side BC, $Q_2 = 1$. The substitution of this value, together with (17.4), into (17.1) leads to a first-order equation for Q_1. The solution of this equation satisfying the isotropy condition (14.5c) at the center is

(17.5) $Q_1 = 1 - P\xi^2 + (\ddot{\phi}/12)(2\xi^2 - \xi^3)$

Finally, the condition (14.5a) that the moment vanish at the simply supported edge enables us to find $\ddot{\phi}$ for any load P. The complete solution may then be written

(17.6)
$$Q_1 = (1 - \xi)[1 + \xi + (P - 1)\xi^2], \quad Q_2 = 1$$
$$W = 12(1 - \xi)\int_0^\tau \int_0^\tau (P - 1)(d\tau)^2$$

Now, in addition to the equations and boundary conditions which determine the above solution, certain inequalities exist which must also be satisfied. These consist of (17.3b) and the condition

(17.7) $0 \leq Q_1 \leq 1$

expressing the fact that the stress profile is on the finite side BC rather than on BC extended. It is readily verified that for any fixed value of P, $Q_1(\xi)$ is equal to 1 at $\xi = 0$ and has a horizontal tangent there, whereas the function vanishes at $\xi = 1$. Further, $Q_1(\xi)$ is a cubic. Therefore, it is evident that necessary and sufficient conditions for the function to satisfy (17.7) are

(17.8) $Q_1''(0) \leq 0, \quad Q_1'(1) \leq 0$

Substitution of the first Eq. (17.6) into (17.8) then shows that P must satisfy

(17.9) $-1 \leq P \leq 2$

The left-hand inequality is always satisfied, and since P is assumed to be monotonically nonincreasing, it is sufficient to satisfy the right-hand inequality at $\tau = 0$.

Obviously a necessary condition that (17.3b) be valid is that P be initially greater than 1. However, the solution may continue to be valid after P has dropped below 1. Indeed, from (17.6), the inequality (17.3b) is

$$(17.10) \qquad \dot{W}' = -12 \int_0^\tau (P - 1)\, d\tau \leq 0$$

Since $P - 1$ is a monotonically nonincreasing function of time which eventually tends to -1, it is evident that (17.10) will be valid for all $0 \leq \tau \leq \tau_2$, where τ_2 is the unique positive solution of

$$(17.11) \qquad \int_0^{\tau_2} [P(\tau) - 1]\, d\tau = 0$$

Further, since $P(\tau_2) < 1$, the load is then below the collapse load and there will be no further motion.

As an example, let the pressure be given by

$$(17.12) \qquad P = Be^{-\beta\tau}$$

In view of (17.9), (17.10), and the definition of the collapse load, the solution (17.6) will be valid only if

$$(17.13) \qquad 1 \leq B \leq 2$$

Therefore, the complete solution for all B satisfying (17.13) is

$$(17.14) \qquad Q_1 = (1 - \xi)[1 + \xi + (Be^{-\beta\tau} - 1)\xi^2], \qquad Q_2 = 1$$

$$\beta^2 W = 12(1 - \xi)[B\beta\tau - \beta^2\tau^2/2 - B(1 - e^{-\beta\tau})], \qquad 0 \leq \tau \leq \tau_2$$

where τ_2 is determined by

$$(17.15) \qquad B = \beta\tau_2(1 - e^{-\beta\tau_2})^{-1}$$

The relation between B and τ_2 is shown in Fig. 19, along with the time τ_1 at which the applied pressure drops to the collapse load. It is evident that the deformation continues for some time at a load less than $P = 1$. Physically, this is because the motion must continue until the kinetic energy imparted by the excess load has been absorbed in plastic deformation. The deformation at the center of the plate at τ_1 and τ_2 is shown in Fig. 20.

If B is greater than 2, then the solution (17.6) predicts values of Q_1 greater than 1 near the center of the plate. It follows that the original hypothesis for the stress profile is not valid in this case, but must be replaced by the type of profile used in Section 16. The details become somewhat more complicated and will not be considered here. A problem similar to this was treated in [17.1].

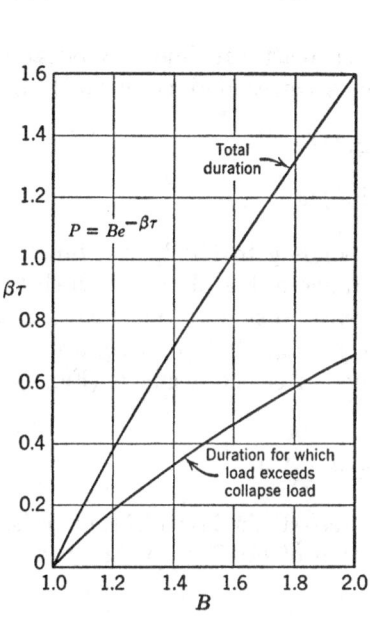

Fig. 19. Duration of plastic deformation for circular plate.

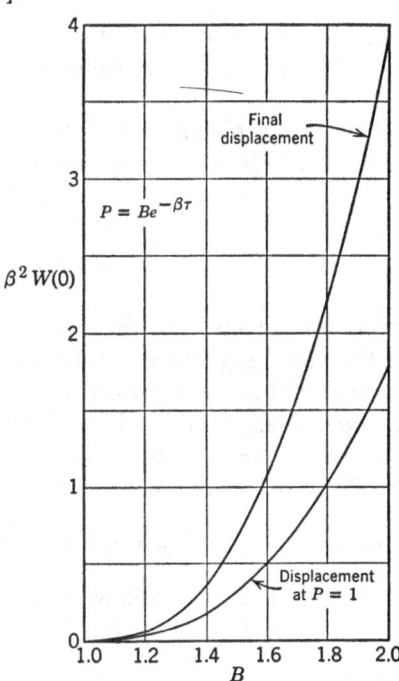

Fig. 20. Displacement of plate center as a function of initial load.

18. Application of principle of minimum potential energy

The minimum principles described in Chapter 4 present potentially powerful tools for the solution of plasticity problems. However, with the exception of the limit analysis theorems, practically nothing has been done by way of application. The only application of the rate principles known to the author is a solution of the thick-walled tube problem by Jung [18.1]. Other than his own applications to the cylindrical shell problems described in [6.1], [9.2], and [18.2] the author knows of no published application of the finite principles proved in Section 12.

Despite this lack of application, it seems worthwhile to develop a simple example in order to illustrate some of the distinctive problems encountered which are not found in elasticity. Let us, then, consider a plate made of a rigid-linear hardening material with isotropic hardening. We assume a displacement function

$$(18.1) \qquad cW = A(1 - \xi^2) + B(1 - \xi^4)$$

which satisfies the boundary conditions on displacement. The constants A and B are to be determined so as to minimize the potential energy (12.7).

The first problem is to determine which plastic regime is applicable. In view of (14.3b), the generalized strains corresponding to (18.1) are

$$(18.2) \qquad \begin{aligned} q_1 &= 2(A + 6B\xi^2)/c \\ q_2 &= 2(A + 2B\xi^2)/c \end{aligned}$$

Since the direction of the vector corresponding to (18.2) is a function of the variable ξ, there will be no finite range of the plate for which the vector will have a constant direction; hence no side of the yield frame is applicable. As a first hypothesis let us assume that q_1 and q_2 are both positive throughout the plate so that the correct corner is B (Fig. 13). At B, the yield function f is given by

$$(18.3) \qquad f = Q_1 = Q_2 = c(q_1 + q_2) + 1$$

where the last step follows from the stress-strain law (9.11c) for isotropic hardening. Finally, the substitution of (18.2) into (18.3) yields

$$(18.4) \qquad f = 4A + 16B\xi^2 + 1$$

In proving (12.7) for the potential energy, the assumption was made that the energy density was equal to $Q_i q_i$, whereas with the present choice of variables it is actually equal to $(Yh^2/a)Q_i q_i$. With the appropriate modification of (12.7) and the use of dimensionless quantities P and W defined by (14.2) and (14.3), the potential energy is proportional to

$$(18.5) \qquad \Pi = \pi a^2 \left[(1/c) \int_0^1 (f^2 - 1)\xi \, d\xi - 12P \int_0^1 W\xi \, d\xi \right]$$

Use of (18.1) and (18.4) then yields

$$(18.6) \quad \Pi = (\pi a^2/c) \left\{ \int_0^1 [(4A + 16B\xi^2 + 1)^2 - 1]\xi \, d\xi \right.$$

$$\left. - 12P \int_0^1 [A(1 - \xi^2) + B(1 - \xi^4)]\xi \, d\xi \right\}$$

If A and B are computed in the usual manner by setting $\partial\Pi/\partial A = \partial\Pi/\partial B = 0$, the results

(18.7) $\qquad\qquad A = (3P - 2)/8, \qquad B = -3P/32$

are obtained. However, when these values are substituted back into (18.2), we find that the strains are

(18.8)
$$q_1 = \left(\frac{1}{8c}\right)\left(6P - 4 - 9P\xi^2\right)$$

$$q_2 = \left(\frac{1}{8c}\right)\left(6P - 4 - 3P\xi^2\right)$$

Now, it will be recalled that the preceding results were obtained under the hypothesis that q_1 and q_2 were both nonnegative throughout the plate, but it follows from (18.8) that q_1 is, in fact, less than 0 near $\xi = 1$. Therefore, we have not obtained a proper minimum for the potential energy, and our solution is meaningless.

We can overcome this difficulty in two ways. On the one hand, we can adopt an alternative hypothesis. For example, we could assume that q_2 was positive but that q_1 was negative for $0 \leq \xi < y$ and positive for $y < \xi \leq 1$. In this case y would be determined from the condition that $q_1(y) = 0$.

In the present exposition we shall adopt a different approach. We shall determine one of the constants so as to enforce the condition that $q_1 \geq 0$. Thus, we require that

(18.9) $\qquad\qquad q_1(1) = (2A + 12B)/c = 0$

or

(18.10) $\qquad\qquad A = -6B$

Therefore, (18.6) is to be replaced by

(18.11) $\quad \Pi = (\pi a^2/c)\left\{ \int_0^1 [(-24B + 16B\xi^2 + 1)^2 - 1]\xi\, d\xi \right.$

$$\left. - 12P\int_0^1 [-6B(1 - \xi^2) + B(1 - \xi^4)]\xi\, d\xi \right\}$$

Determining B so as to minimize (18.11), we obtain

(18.12) $\qquad\qquad B = -\tfrac{3}{416}(7P - 8)$

Further, since the strains are now given by

(18.13) $\qquad cq_1 = -12B(1 - \xi^2), \qquad cq_2 = -4B(3 - \xi^2)$

it is evident that these will both be positive provided

(18.14) $\qquad\qquad\qquad P \geq \tfrac{8}{7}$

Finally, the displacement is given by

(18.15) $\qquad cW = \tfrac{3}{416}(7P - 8)(1 - \xi^2)(5 - \xi^2)$

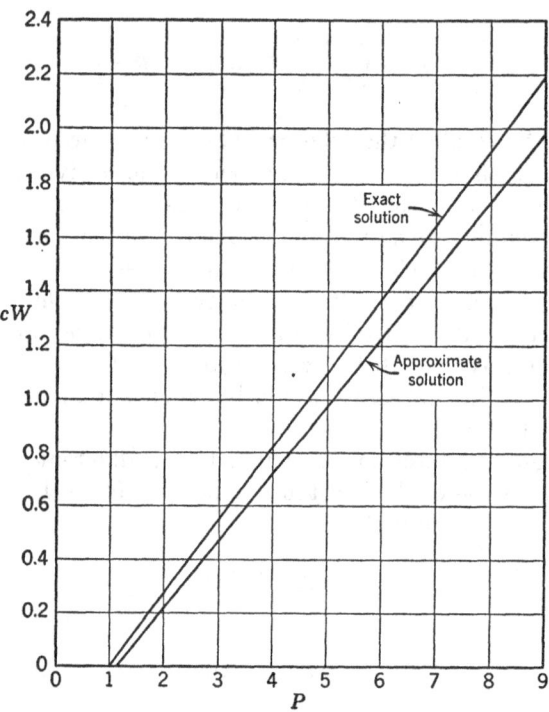

Fig. 21. Displacement at plate center computed by potential energy approximation.

Figure 21 shows the displacement at the center of the plate as a function of the load P. The approximate value computed from (18.15) is compared with the exact value found in Section 16 and is seen to be in reasonably good agreement.

$\bigcirc$THER PROBLEMS

19. Circular cylindrical shell

The development of circular cylindrical shell theory and application has paralleled that of circular plates to quite an extent. Thus shells under radial pressure only have been solved for rigid-perfectly plastic materials [19.1], elastic-plastic materials [19.2], and linear hardening materials [19.3]. Dynamic loading has also been investigated [19.4, 19.5, 19.6, 19.7]. Consideration has also been given to infinite [19.8] and finite [19.9] shells under a concentrated ring load or a band of pressure. Shells under combined radial pressure and end load have also been investigated [19.1, 19.10]. A beginning attempt has been made to extend results to the general axially symmetric shell [19.11].

The essential difference between plastic shell and plate theory is the representation of the yield condition in terms of suitable generalized co-ordinates. Let us, for simplicity of exposition, consider the case of a shell under radial pressure only. In this case the only stress resultants which do work are the circumferential direct stress N_ϕ and the axial bending moment M_x. Therefore, we choose as dimensionless variables

$$Q_1 = N_\phi/2Yh, \qquad Q_2 = M_x/Yh^2$$

(19.1)

$$q_1 = e_\phi, \qquad q_2 = h\kappa_x/2$$

The axial direct stress does no work, since

(19.2) $$N_x = 0$$

whereas changes of circumferential curvature are neglected so that the moment M_ϕ does no work.

Assuming that the shell material yields according to the Tresca criterion of maximum shearing stress,

(19.3) $$\max[\,|\sigma_x|,\ |\sigma_\phi|,\ |\sigma_x - \sigma_\phi|\,] - Y \le 0$$

we observe first that in order for a section of the shell to be plastic at least one of the six inequalities (19.3) must be an equality throughout the section. Further, admissible stress distributions must satisfy (19.3). However, there is no necessity for the same equality in (19.3) to hold throughout the section.

As an example, we may consider those stress distributions which produce the full bending moment M_x. The only admissible σ_x distribution satisfying (19.3) is shown in Fig. 22a. However, the σ_ϕ

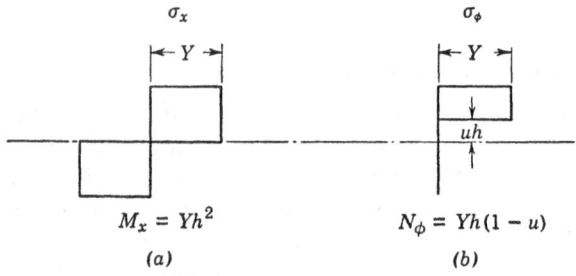

σ_x σ_ϕ

$M_x = Yh^2$ $N_\phi = Yh(1-u)$

(a) (b)

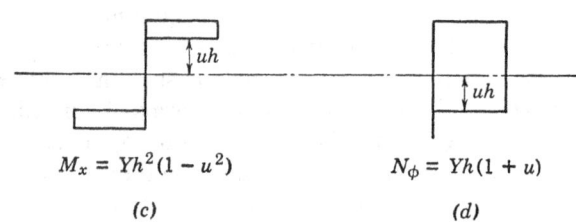

$M_x = Yh^2(1-u^2)$ $N_\phi = Yh(1+u)$

(c) (d)

Fig. 22. Fully plastic stress distributions across shell section. (a) σ_x for full moment. (b) σ_ϕ for full moment. (c) σ_x for less than full moment. (d) σ_ϕ for less than full moment.

distribution shown in Fig. 22b is valid for any value of the parameter u. It follows that the line segment

(19.4a) $$Q_1 = 1, \qquad 0 \le Q_2 \le \tfrac{1}{2}$$

is part of the yield curve.

Values of Q_2 greater than $\tfrac{1}{2}$ can be obtained only by decreasing the moment. The most efficient (in the sense of preserving the maximum allowable Q_1) way of doing this is shown in Fig. 22c and d. The corresponding stresses are

(19.4b) $Q_1 = 1 - u^2, \qquad Q_2 = (1 + u)/2, \qquad 0 \le u \le 1$

Since similar results hold in each quadrant, the complete yield curve is obtained by symmetry; it is shown as the solid curve in Fig. 23.

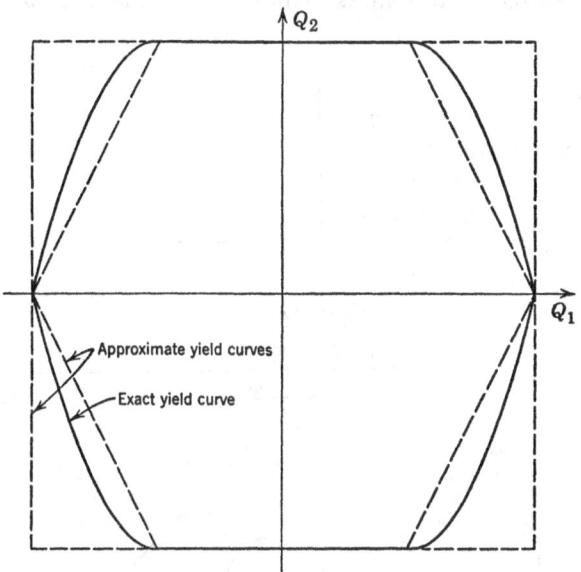

Fig. 23. Yield curve for circular cylindrical shell.

Strictly speaking, we have shown only that no point of the yield curve lies inside the curve in Fig. 23. This follows from the first principle of limit analysis, since we have actually constructed only a statically admissible stress field, and hence a lower bound. However, in this simple example it is evident that we have exhausted all possible lower bounds and thus have the true yield condition.

An alternative method of obtaining the curve in Fig. 23 is to consider first the admissible directions of the strain rate vector, regarding the strains as fully plastic. If the strain rates are prescribed, the stresses are at least partially determined from the flow rule. For example, if the strain rate vector leads to any point in the interior of the first quadrant of an $\dot{\epsilon}_x$, $\dot{\epsilon}_\phi$ space, the stress point must be at the corner $\sigma_1 = \sigma_2 = Y$ of the yield curve. However, if the strain rate vector is along the positive $\dot{\epsilon}_x$ axis, the stress point can lie anywhere on the side $\sigma_x = Y$.

Thus the flow law represents a mapping between the yield curve and a strain rate space, as indicated in Fig. 24.

Now, the strain rates at any point in a shell section are given by

$$(19.5) \qquad \dot{\epsilon}_x = \dot{e}_x + z\dot{\kappa}_x, \qquad \dot{\epsilon}_\phi = \dot{e}_\phi$$

where e_x, e_ϕ, and κ_x are the corresponding extensions and curvature of the middle surface. It follows that if the strain rates at the middle

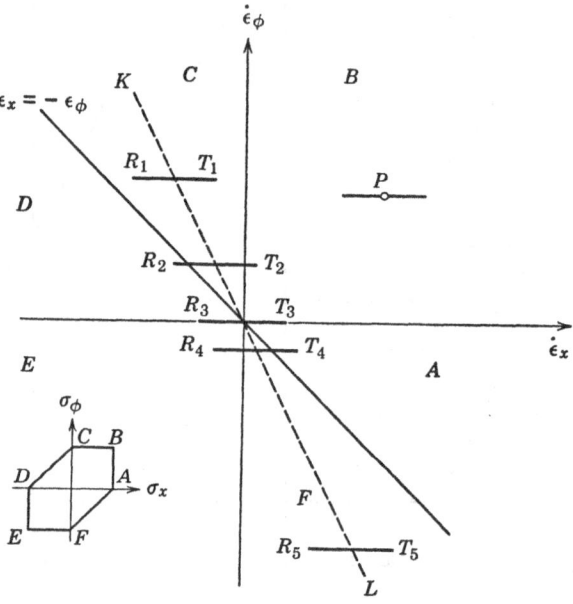

Fig. 24. Strain rate approach to yield curve for circular cylindrical shell.

surface are represented by a point P in Fig. 24, then the strains across the cross section will be represented by a line segment of length $2h\dot{\kappa}_x$ which is centered at P and parallel to the $\dot{\epsilon}_x$ axis. For the particular point P shown in Fig. 24, it then follows that the stress at all points of the cross section must be $\sigma_x = \sigma_y = Y$. However, this leads to a non-zero value for the axial stress resultant N_x, hence it is not acceptable for the present problem. In fact, it is evident that the requirement $N_x = 0$ restricts the strain rate point for the middle surface to lie on the dashed line KL in Fig. 24.

We can now distinguish five categories, as illustrated by the line segments $R_\alpha T_\alpha$ in Fig. 24. $R_1 T_1$ implies that $\sigma_x = 0$, $\sigma_\phi = Y$ across the section, and leads to the stress resultants $Q_1 = 0$, $Q_2 = 1$. The segment R_2T_2 is equivalent (assuming $\kappa_x > 0$) to a stress distribution of the form

$$\sigma_x = -Y, \qquad \sigma_\phi = 0, \qquad -h \leq z \leq -uh$$

(19.6)
$$\sigma_x = 0, \qquad \sigma_\phi = Y, \qquad -uh \leq z \leq uh$$

$$\sigma_x = Y, \qquad \sigma_\phi = Y, \qquad uh \leq z \leq h$$

This is precisely the stress distribution shown in Fig. 22c and d and hence it leads again to (19.4b). In a similar fashion we can show that the stress distribution shown in Fig. 22a and b is compatible with the line segment R_3T_3, and the remaining two segments lead to distributions symmetric with those of Fig. 22. Thus, by this approach we are again led to the yield curve of Fig. 23.

Viewed by itself, this second method is based upon kinematically admissible strain rates. Hence, it follows from the second principle of limit analysis that the resulting yield curve is an "upper bound" in that the true yield curve cannot lie outside it. Since we have exhausted all possible upper bounds, we have obtained the true yield curve. The final proof of this last statement is the fact that each of the two methods used has led to the same yield curve.

An interesting feature of this problem is that although we started with a piecewise linear yield function in terms of the stresses, the resulting function of the stress resultants is partially nonlinear. If we wish to use piecewise linear theory, this curve must now be approximated. The dashed polygons in Fig. 23 indicate two approximations which have been used in the literature. The details of applications are quite similar to those encountered in plate theory and will not be entered into here. The interested reader is referred to the references given at the beginning of the section.

20. Plane strain and plane stress

The mathematical theory of plane strain for a rigid-perfectly plastic material is one of the oldest branches of plasticity. Considerable knowledge of the governing differential equations had been accumulated prior to 1930. Later, the application of this theory to the solution of practical boundary value problems was investigated. Quite extensive accounts of both theory and application may be found in [0.3] and [0.4] and will not be repeated here.

Since 1950, many further applications have been published. Without trying very hard, the author was able to collect a list of over 70 such solutions, and it would obviously be absurd to attempt to describe, or even list, all of these. Therefore, we shall content ourselves with very brief reviews of a few of the solutions which appear to be particularly significant.

On the theoretical side, the only advances in plane strain have been towards improved methods of solution. Among these may be mentioned a graphical construction due to Prager [20.1, 20.2] and a somewhat similar hodograph method developed by Green [20.3].

With regard to plane stress, the theory has been developed to a point comparable with plane strain. A rather complete survey of two-dimensional problems has been published by Geiringer [20.4], including an extensive bibliography. From the viewpoint of obtaining solutions, an important development has been the use of the Tresca yield condition with its associated flow rule. Many of the plane stress solutions mentioned in the following text have been obtained under this hypothesis.

Problems involving notched bars have been among the most popular in the recent literature. Tension under conditions of plane strain has been treated by Lee and others [20.5, 20.6, 20.7]. Closely related problems of necking have been investigated by Hill [20.8], Onat and Prager [20.9], and Thomas [20.10]. Green has discussed the bending of notched bars [20.11], and has also treated the bending of beams as a two-dimensional problem [20.12]. Gaydon [20.13], Craggs [20.14], Horne [20.15], and Onat and Shield [20.16] have also worked on bending problems.

Upper and lower bounds on the collapse loads of plane slabs with cutouts have been found by Gaydon [20.17, 20.18] and Hodge [20.19]. The effect of reinforcements on the cutouts has also been considered [20.19, 20.20, 20.21].

All of the above papers are concerned with rigid-plastic materials, or with the collapse loads of elastic-plastic materials. The complete solutions available are, for the most part, restricted to rotationally symmetric problems. In addition to uncounted papers on thick-walled tubes, these include a few solutions to thin sheet problems where the change in thickness has been accounted for [8.2, 8.3, 20.22].

Problems of metal forming in plane strain were quite thoroughly investigated prior to 1950, and numerous examples of extrusion, drawing, and rolling may be found in [0.3] and [0.4]. More recently, the process of chip formation has been treated as a plasticity problem by Lee and Shaffer [20.23, 20.24, 20.25].

As a final example, we consider the problem of a semi-infinite plastic solid indented by a rigid punch. This was one of the earliest problems for which a velocity solution was found, but only recently have Shield and Drucker [20.26] obtained a lower bound on the indentation load by considering discontinuous stress fields. These results have been extended to certain finite plastic solids by Shield [20.27] and Ross [20.28]. The related problem of a three-dimensional punch was treated in [20.26] and also by Levin [20.29].

21. Beams, bars, and rods

The limit analysis of beams and frames has been extensively developed during the past few years. This work has been done under the assumptions that the influence of shear force and axial force on yielding can be neglected, so that the beam characteristics can be described by a moment-curvature relation. This latter is assumed to be similar to the stress-strain curves shown in Fig. 1b, d, f, so that strain hardening can be neglected. A comprehensive account of the work done under these assumptions has been given by Symonds and Neal [21.1]. Although many papers have been published since then, the main lines of development may be regarded as well established, and the subject will not be treated further in the present article.

We shall, however, comment briefly on some of the extensions of this work. Hoff [21.2] has computed a particular indeterminate beam of an elastic-linear hardening material, and has shown that the complete solution is very well approximated by the theory used in limit analysis. Numerous authors [20.12, 20.13, 20.14, 20.15, 20.16] have investigated the effect of shear on the yielding of beams, by regarding the beam as a two-dimensional continuum in a state of plane strain or plane stress. Onat and Prager [21.3] have considered the effect of axial forces on the collapse loads of frames. Closely related to this latter question are the collapse loads of arches and rings where axial forces must be important. Papers on this topic have been written by Hendry [21.4], Onat and Prager [21.5], and Hwang [21.6]. Some elastic-plastic solutions for curved beams have been given by Swida [21.7, 21.8, 21.9], Ohno [21.10], Phillips [21.11], and Shaffer and House [21.12].

The dynamic problem for a rigid-plastic material which was discussed in Section 17 was first solved for an infinite beam. Lee and Symonds [21.13] carried out this first investigation, and Symonds and others have solved many additional dynamic beam problems [21.14, 21.15, 21.16]. Similar problems have been handled by Conroy [21.17], Bleich and Salvadori [21.18], Parkes [21.19], and Mentel [21.20].

The basic theory for the torsion of elastic-perfectly plastic bars has been known for some time. It is fully described in [0.3] or [0.4]. Recent contributions include an extension to work-hardening materials by Onat [21.21], a discussion of the finite torsion problem by Seth [21.22], and an extension to anisotropic materials by Hill [21.23].

Finally, various combinations of bending, torsion, and tension of bars have been considered. In particular, Brush, Sidebottom, and Smith [21.24], Barrett [21.25], and Frankland and Roach [21.26] have investigated combined tension and bending. Papers on combined tension and torsion have been written by Mii [21.27] and Gaydon [21.28]. The combination of bending and torsion has been considered by Hill and Siebel [21.29], Mii [21.30], and Steele [21.31].

The computational problem in the plastic treatment of complex frames has received some attention. Foulkes [21.32] has reduced the minimum weight design problem to one of linear programming. Kron [21.33] has extended his method of "tearing" to include plastic structures.

22. Miscellaneous problems

Closely related to the beam problems discussed in the previous section are the problems of thin-walled tubes. The combination of bending and twisting of thin-walled tubes has been investigated by Hill and Siebel [22.1, 22.2] and by Onat and Shield [22.3].

The problem of a thick-walled tube under combined tension and torsion has been treated by Crossland and Hill [22.4]. Christopherson and Higginson [22.5] have looked into the thick-walled cylinder under pressure when the cylinder is very short.

The general class of axially symmetric problems is of considerable interest in a variety of applications. Theoretical papers on the subject have been written by Symonds [22.6] and Jung [22.7]. One of the most important theoretical papers is Shield's treatment [22.8] of the axially symmetric problem by means of Tresca's yield condition.

Thomsen and his associates [22.9, 22.10, 22.11, 22.12] have performed some valuable experimental and theoretical work on the extrusion of cylindrical billets. Shield [22.13] has found a solution for a theoretical problem which closely approximates wire drawing. The problem of tube drawing has been investigated by Swift [22.14]. The deep drawing of metal cups has been the subject of papers by Hill [22.15], Chung and Swift [22.16], and Yamada [22.17]. Hill [22.18], Ross and Prager [22.19], and Weil and Newmark [22.20] have considered the bulging of a metal diaphragm.

CHAPTER 7

Russian Contributions

23. General remarks

For a period of about 10 years starting approximately in 1935, it may truthfully be said that Russia was the center of activity in research in mathematical plasticity. Since that time there has been a tremendous increase in interest in the United States and Great Britain, so that leadership has apparently passed to these countries. However, Russian scientists have continued to do significant work in the field, and it is the purpose of the present chapter to review that work.

In general terms, there are some rather striking differences between the Russian approach to plasticity and that found in the English-speaking countries. On the whole, Russian scientists have based their work upon a deformation theory of plasticity, as opposed to the flow theories developed in Chapters 1 and 2. This is in spite of the fact that the Russian author Ilyushin [24.7] was one of the first to clearly formulate the difference between the two laws. Although it might therefore seem as if Russian contributions could be of little value in relation to the more realistic flow laws, this would not be a just evaluation. As mentioned in Section 3 and illustrated in Chapter 3, there are numerous problems where a deformation law predicts results sufficiently close to those of a flow law. Although none of the Russian authors justify their use of a deformation law by an explicit statement to this effect, it is still a valid argument. There is much to be said in favor of a careful analysis of the Russian problem solutions from a viewpoint of deciding whether or not they represent such useful approximations.

A second possible value of the Russian work is that it may contain useful hints for obtaining solutions by flow theories. In general, a plasticity problem is nonlinear regardless of the theory upon which it is based. So little is known about the solution of nonlinear problems that we can ill afford to neglect any possible aid in obtaining a solution. Such aid may consist simply of providing a reasonable "first guess" in an iterative or relaxation type of numerical procedure, or it may consist in suggesting a novel and fruitful way of using advanced mathematical techniques.

Another basic difference of the Russian viewpoint is that it appears to have very little regard for experimental results. Prior to 1949, there were few, if any, references to Russian experimental efforts, nor was much effort made to compare theoretical results with experimental results from other countries. Although this situation has changed slightly during the past 6 years, the experimentalist still appears to play a minor role compared to his importance in other countries.

On the positive side, the Russians have not hesitated to apply advanced mathematics to the solution of plasticity problems. The complete elastic-plastic solution of a certain type of problem in plane strain by Galin [24.42] (see also the account of Galin's work in [0.4], Sec. 31) is an example whose elegance has not been approached outside of that country. Further, they have not been content with purely formal mathematical indications of a solution, but have made free use of numerical techniques in order to obtain specific answers.

Such then is the general state of Russian plasticity. In view of the relative unavailability of many of the Russian publications, it was felt advisable to make a rather exhaustive survey of the Russian literature, as opposed to the more superficial listing of references in the previous sections. Such a survey was previously done by W. Prager in a Brown University Report to the Air Matériel Command in 1950. This report was originally classified as confidential, and although it has since been declassified, it is no longer available. Rather than attempt to digest or rearrange this excellent work, the author has obtained permission from Professor Prager to quote the relevant portions of it in full. This, rather than any significant change of emphasis, is the reason for dividing the Russian contributions into two chronological periods in the remainder of this chapter. Within each section, the plan has been to give a complete listing of Russian papers in the bibliography, but to refer in the text only to those deemed the most significant.

24. Contributions up to 1949 *

The Bibliography {for this section} contains {74} references and is believed to represent a fairly complete listing of Russian papers on the mathematical theory of plasticity up to about the middle of 1949. The references of this bibliography are grouped as follows:

I. General theory
II. Special problems
  (a) Torsion
  (b) Thick-walled spheres and cylinders
  (c) Rotating disks
  (d) Plane stress and plane strain
  (e) Plates and shells
  (f) Rolling, drawing, and extruding processes
  (g) Problems of contact stress and indentation
  (h) Structural stability in the plastic range ...

The rest of this {section} will be devoted to brief comments on those Russian papers on the mathematical theory of plasticity which, in the opinion of the present writer {Prager}, represent outstanding contributions to this theory. Each paragraph will be labeled by the reference number of the paper which it discusses.

[24.2]. This paper contains an excellent discussion, in the case of a continuum, of the principle which is the basis of the method of "limit design"[1] and hints how this principle might be used to obtain upper and lower bounds for the safety factor [13.3]. The method of limit design was first suggested by N. C. Kist[2] in his inaugural lecture at the Technical University of Delft (Oct. 2, 1917). It was widely discussed in German literature of the thirties[3] and has recently attracted attention in Great Britian[4] and the United States[1] [13.3]. It appears that this

---

* This entire section, together with the corresponding items in the bibliography, represents a complete transcription of the relevant portions of Prager's report referred to in the previous section, except that reference and footnote numbers have been changed to fit the format of the present work. Omissions are indicated by three dots ... and inclusions or changes by material in braces { }. References to non-Russian work prior to 1950 are listed as footnotes in order to preserve the general aim of the present article and at the same time list those sources which Prager considered relevant.

[1] See, for instance, J. A. Van den Broek, *Theory of Limit Design*, John Wiley & Sons, New York, 1948.

[2] *Eisenbau*, **11**, 425, 1920.

[3] See, for instance, H. Bleich, *Bauingenieur*, **13**, 261–267, 1932; E. Melan, *Sitzber. Akad. Wiss. Wien*, **145**, 195–218, 1936.

[4] See, for instance, a series of reports on the behavior of welded rigid frame structures by J. F. Baker, J. W. Roderick, and others to the Institute of Welding.

design procedure was incorporated into Russian building codes long before any such step was contemplated in other countries.

[24.7]. Although the stress analyses of plates and shells presented in this paper are based on a finite stress-strain law, the methods used are of general interest.

[24.8]. This paper contains a good discussion of "flow" and "deformation" theories and raises, for the first time, the question under what conditions of loading the predictions of the two theories agree.

[24.15]. The first Russian paper devoted to the mathematical problems arising in the discussion of statically determinate stress distributions in perfectly plastic solids. The method of characteristics is used and the problems are simultaneously discussed in the physical plane (x, y) and the plane of the characteristic parameters (ξ, η). The concept of limiting line (called "line of rupture") is introduced and the behavior of the stress distribution in the neighborhood of a limiting line is discussed.

[24.18]. Minimum principles are established for the velocity field in a perfectly plastic solid under the assumption that the boundary conditions are such as to assure plastic behavior *throughout* the body. These principles are used to establish uniqueness of the stresses to within an arbitrary hydrostatic pressure. The relation of these minimum principles to the maximum principles of M. A. Sadowsky [10.2] and R. Hill [10.3] has been discussed by the latter.

[24.22]. This book is the most exhaustive text on the theory of perfectly plastic solids available to date. In the present writer's opinion, the statically determinate stress distributions are emphasized too much and the interplay of the stress and velocity equations is not brought out as much as it should be. In spite of this, Sokolovsky's book constitutes a definite advance beyond the earlier surveys {by Nadai,[5] Odqvist,[6] and Geiringer and Prager [7]}, and also beyond earlier Russian surveys [24.16, 24.19].

[24.31]. This paper contains the first nontrivial exact solution of a problem in elastic-plastic torsion. The inverse method used by Sokolovsky has been further studied by L. A. Galin [24.26]; another example has been given by R. von Mises.[8] {An account of this method may be found in [0.4], Sec. 11.}

[24.42]. The author remarked that the stress function for the fully plastic stress distribution in the neighborhood of a circular hole is bi-

---

[5] *Handbuch der Physik*, vol. 6, Springer, Berlin, Chap. 6, 1928.

[6] *Plasticitetsteori med tillaempiger* (Swedish), Stockholm, 1934.

[7] *Ergeb. exakt. Naturw.*, **13**, 310–363, 1934.

[8] *Reissner Anniversary Volume*, Ann Arbor, 415–429, 1949.

harmonic as is the elastic stress function. Using this fact (which W. Prager [9] had established in a different manner), the author succeeds in analyzing the plane elastic-plastic strain in an infinite plate with a circular hole when the state of stress at infinity is given and the hole is subjected to a uniform pressure such that the elastic-plastic boundary encloses the hole. This constitutes the first exact solution of a non-trivial problem in plane elastic-plastic strain. {An account of this method may be found in [0.4], Sec. 31.}

[24.45]. The method which was used by Galin in the paper discussed in the preceding paragraph depended on the fact that, in the particular case under consideration, the stress function was biharmonic in both the plastic and elastic regions. Obviously, a method depending on this coincidence is not of general interest. Parasyuk [24.45] succeeded in adapting Muskhelishvili's method in elasticity to a wider class of problems in plane elastic-plastic strain.

[24.49, 24.50, 24.51, 24.52]. These papers represent the first general investigations of statically determinate problems of plane stress in perfectly plastic solids. Contrary to what is the case in problems of plane strain, the basic equations may change from the hyperbolic to the elliptic type, and the characteristics in the hyperbolic field do not coincide with the lines of maximum shearing stress.

[24.60]. The author develops an adequate approximate theory for the forming of thin-walled circular tubes of varying diameter by drawing or extruding. For the theory to be applicable, the ratio of wall thickness to diameter must be much smaller than is the case for normal ammunition. It is possible, however, that the "tubes" considered in the paper are rocket or torpedo shells.

[24.62, 24.63, 24.64, 24.65]. The papers represent the first attempt at applying the full mathematical theory of plasticity to the rolling problem. (Previous papers on this subject always combined some equations of the theory of plasticity with *ad hoc* assumptions regarding the distribution of stress or strain.) The work suffers from the fact that the author does not realize the necessity to reconcile the stress field with the boundary conditions on velocities (see, for instance, R. Hill and S. J. Tupper [10]) {also [0.3] or [0.4].}

[24.70]. This paper represents the first rigorous attempt at extending Kármán's double modulus theory of the plastic buckling of columns to the plastic buckling of plates. The results are questionable, however, because the author uses a finite stress-strain law. The corresponding theory for an incremental stress-strain law was established by G. H.

---

[9] *Rev. Math. Un. Interbalk.*, **2**, 45–51, 1938.
[10] *J. Iron Steel Inst. (London)*, **159**, 353–359, Aug. 1948.

Handelman and W. Prager [11] in this country and by I. P. Kuntze [24.74] in Russia.

25. Contributions from 1949 to 1955

The same general scheme of presentation will be used here as in the previous section, except that the last few categories will be grouped together. The list includes a few papers from the other Communist countries, and a few papers with publication dates prior to 1949 which were not immediately available in this country.

[25.1]. Only the last quarter of this book is concerned with plasticity. An excellent survey of Russian work at the time is given, but little mention is made of any other contributions.

[25.2]. Three methods of approximate solution of deformation law plasticity problems are given. Each method is based upon a related elastic problem.

[25.8, 25.9]. A finite-strain theory for small deformations is developed. The yield condition depends upon the first two (not the third) stress invariants.

[25.15, 25.11, 25.21, 25.22, 25.33, 25.16]. The first of these papers represents an address delivered before the general session of the Technological Section of the Academy of Science of the U.S.S.R. in June 1949. It is primarily concerned with the argument between mathematicians and engineers on the one hand and physicists and metallurgists on the other. Ilyushin defends the viewpoint of the engineer who has no time to wait until the physicist can build up a theory from first principles— and even if the theory were built it would probably be too complex to help the engineer. The next four papers [25.11, 25.21, 25.22, 25.33] are attacks on Ilyushin's paper which appeared shortly thereafter. Some of these raise legitimate scientific points of argument; others are personal and almost vitriolic. The final paper [25.16] is Ilyushin's answer to his various attackers. The papers are interesting in that they show that (at least in this scientific field and at that time) considerable basic controversy between scientists was allowed and was even printed.

[25.17]. This is a nonmathematical review of contributions to plasticity theory from Moscow State University. There are 45 recent references, but little correlation between the references and the review.

[25.19]. This is a book on limit analysis, addressed to practicing engineers. It is based on the Russian building code and contains sections on specific structural materials.

[25.20]. One of the few Russian experimental papers, it compares

[11] *Technical Note No. 1530*, NACA, 1948.

the yield stresses in tension and shear of various materials, finding a tremendous range in value for their ratio.

[25.24]. An extreme case of the insularity of some of the Russian work, this paper is concerned with the discrepancy between the Mises and Tresca yield criteria. It re-establishes the result first found by Mises [1] in 1913 that the maximum difference is less than 16%.

[25.30]. A Galerkin-type method of solution is adapted to elastic-plastic materials satisfying a deformation law.

[25.32]. Rolling of an anisotropic sheet is discussed. Good agreement is found between theoretical and experimental results.

[25.36]. This is a formulation of a theory for a rigid-perfectly plastic material, satisfying a flow law. Incompressibility is not assumed. The yield condition depends arbitrarily upon the first and second stress invariants, but not the third. With these assumptions Sokolovsky discusses several transformations of the two-dimensional equations, whether of elliptic or hyperbolic type. A general method of solution by means of trigonometric series is given, analogous to that used by Chaplygin in the theory of plane gas flows.

[25.40, 25.41]. The paper by Galin [25.41] is a systematic study of the inverse method of solving the elastic-plastic torsion problem, first presented by Sokolovsky [24.31]. Bulygin's paper [25.40] is an example of this method.

[25.42]. This paper is concerned with the existence proof for the torsion problem of a certain class of elastic-plastic materials.

[25.48]. In addition to the usual solution to the thick-walled tube problem, a dynamic analysis is given for the case when the pressure is applied as an impulse.

[25.51]. A finite-strain solution is presented for various tube and shell problems.

[25.63, 25.64]. These papers represent further extensions of [24.45] for determining complete elastic-plastic solutions in plane strain. The case of linear hardening according to a deformation theory is considered.

[25.79, 25.80, 25.81]. Deformation law with hardening is used to analyze cylindrical shell problems. In [25.80] a variational principle is used to obtain the equations in the plastic region.

[25.84]. A method of successive approximations based upon elastic solutions is used to solve the elastic-plastic plate problem. Using several advanced mathematical techniques, the author shows that the method converges.

[1] R. von Mises, *Goettinger Nach., math.-phys. Kl. 1913*, 582–592, 1913.

BIBLIOGRAPHY

References are listed in the order in which they first appear, with the exception of references to the Russian literature which are all listed under Chapter 7.

INTRODUCTION

0.1 A. M. Freudenthal, *The Inelastic Behavior of Engineering Materials and Structures*, John Wiley & Sons, New York, 1950.

0.2 A. Nadai, *Theory of Flow and Fracture of Solids*, Vol. 1, McGraw-Hill Book Co., New York, 1950.

0.3 R. Hill, *The Mathematical Theory of Plasticity*, Oxford University Press, London, 1950.

0.4 W. Prager and P. G. Hodge, Jr., *Theory of Perfectly Plastic Solids*, John Wiley & Sons, New York, 1951.

0.5 A. Nadai, *Plasticity, a Mechanics of the Plastic State of Matter*, McGraw-Hill Book Co., New York, 1931.

0.6 H. F. Bohnenblust and P. Duwez, Some properties of a mechanical model of plasticity, *J. Appl. Mech.*, **15**, 222–225, 1948.

CHAP. 1.   THEORY OF PERFECTLY PLASTIC SOLIDS

1.1 W. Prager, The general theory of limit design, *Proc. Eighth Internat. Congr. Appl. Mech.*, **2**, 65–72, 1956.

2.1 D. C. Drucker, Some implications of work hardening and ideal plasticity, *Quart. Appl. Math.*, **7**, 411–418, 1950.

2.2 D. C. Drucker, A more fundamental approach to plastic stress-strain relations, *Proc. First U. S. Nat. Congr. Appl. Mech.*, 487–491, 1952.

2.3 R. von Mises, Mechanik der plastischen Formaenderung von Kristallen, *Z. angew. Math. Mech.*, **8**, 161–185, 1928.

2.4 J. F. W. Bishop and R. Hill, A theory of the plastic distortion of a polycrystalline aggregate under combined stresses, *Phil. Mag.* (7), **42**, 414–427, 1951.

2.5 T. Y. Thomas, Interdependence of the yield condition and the stress-strain relations for plastic flow, *Proc. Nat. Acad. Sci.* (*U. S.*) **40**, 593–597, 1954.

3.1 F. Edelman, On the coincidence of plasticity solutions obtained with incremental and deformation theories, *Proc. First U. S. Nat. Congr. Appl. Mech.*, 493–498, 1952.

3.2 G. H. Handelman and W. H. Warner, Loading paths and the incremental strain law, *J. Math. and Phys.*, **33**, 157–164, 1954.

CHAP. 2. THEORY OF STRAIN-HARDENING PLASTIC SOLIDS

4.1   F. D. Stockton and D. C. Drucker, On fitting mathematical theories of plasticity to experimental results, *J. Colloid Sci.*, **5**, 239–247, 1950.

4.2   A. Phillips, Combined tension and torsion tests for aluminum alloy 2S–O, *J. Appl. Mech.*, **19**, 496–500, 1952.

4.3   A. Phillips and L. Kaechele, Combined stress tests in plasticity, *J. Appl. Mech.*, **23**, 43–48, 1956.

5.1   W. Prager, The theory of plasticity—a survey of recent achievements, *Proc. Inst. Mech. Engrs.*, London, 3–19, 1955.

5.2   P. G. Hodge, Jr., Piecewise linear plasticity, *Proc. Ninth Internat. Congr. Appl. Mech.* (to be publ.).

5.3   P. G. Hodge, Jr., A general theory of piecewise linear plasticity based on maximum shear, *J. Mech. Phys. Solids*, **5**, 242–260, 1957.

6.1   P. G. Hodge, Jr., Piecewise linear isotropic plasticity applied to a circular cylindrical shell with symmetrical radial loading, *J. Franklin Inst.*, **263**, 13–33, 1957.

6.2   P. G. Hodge, Jr., Minimum principles of piecewise linear isotropic plasticity, *J. Rat. Mech. Anal.*, **5**, 917–938, 1956.

6.3   P. G. Hodge, Jr., Discussion of Ref. [16.1]. *J. Appl. Mech.*, **24**, 482–483, 1957.

6.4   J. F. Besseling, A theory of plastic flow for anisotropic hardening in plastic deformation of an initially isotropic material, *Report S410*, Nat. Lucht. Lab. Amsterdam, 1953.

7.1   F. D. Stockton, Experimental evidence of non-linearity in plastic stress-strain relations, *GDAM Report A11-88*, Brown University, 1953.

7.2   B. Budiansky, N. F. Dow, R. W. Peters, and R. P. Shepherd, Experimental studies of polyaxial stress-strain laws of plasticity, *Proc. First U. S. Nat. Congr. Appl. Mech.*, 503–512, 1952.

7.3   S. B. Batdorf and B. Budiansky, A mathematical theory of plasticity based on the concept of slip, *NACA Tech. Note 1871*, 1949.

7.4   P. Cicala, Sobre la teoria de Batdorf y Budiansky de la deformacion plastica, *Rev. Univ. Nac. Cordoba (Arg.)*, **13**, 401–405, 1950.

7.5   R. W. Peters, N. F. Dow, and S. B. Batdorf, Preliminary experiments for testing basic assumptions of plasticity theories, *Proc. Soc. Exp. Stress Analy.*, **7**, 127–140, 1950.

CHAP. 3. PIECEWISE LINEAR PLASTICITY

8.1   W. T. Koiter, On partially plastic thick-walled tubes, *Biezeno Ann. Vol.*, Haarlem, 232–251, 1953.

8.2   W. Prager, On the use of singular yield conditions and associated flow rules, *J. Appl. Mech.*, **20**, 317–320, 1953.

8.3   P. G. Hodge, Jr., On the plastic strains in slabs with cutouts, *J. Appl. Mech.*, **20**, 183–188, 1953.

9.1   J. L. Sanders, Jr., Plastic stress-strain relations based on linear loading functions, *Proc. Second U. S. Nat. Congr. Appl. Mech.*, 455–460, 1955.

9.2   P. G. Hodge, Jr., The theory of piecewise linear isotropic plasticity, in *Deformation and Flow of Solids*, R. Grammel, Ed., J. Springer, Berlin, 147–169, 1956.

9.3   G. N. White, Jr., Application of the theory of perfectly plastic solids to stress analysis of strain hardening solids, *GDAM Report A11-51*, Brown University, 1950.

CHAP. 4. MINIMUM PRINCIPLES OF PLASTICITY

10.1   G. Colonnetti, De l'équilibre des systems élastiques dans lesquels se produisent des déformations plastiques, *J. Math. Pures Appl.* (9), **17**, 233–255, 1938.

10.2   M. A. Sadowsky, A principle of maximum plastic resistance, *J. Appl. Mech.*, **10**, A65–A68, 1943.

10.3   R. Hill, A variational principle of maximum plastic work in classical plasticity, *Quart. J. Mech. Appl. Math.*, **1**, 18–28, 1948.

10.4   R. Hill, A comparative study of some variational principles in the theory of plasticity, *J. Appl. Mech.*, **17**, 64–66, 1950.

10.5   R. Hill, New horizons in the mechanics of solids, *J. Mech. Phys. Solids*, **5**, 66–74, 1956.

10.6   A. H. Philippidis, The general proof of the principle of maximum plastic resistance, *J. Appl. Mech.*, **15**, 241–242, 1948.

10.7   A. Phillips, Variational principles in the theory of finite plastic deformations, *Quart. Appl. Math.*, **7**, 110–114, 1949.

10.8   H. J. Greenberg, Complementary minimum principles for an elastic-plastic material, *Quart. Appl. Math.*, **7**, 85–95, 1949.

10.9   H. J. Greenberg, On the variational principles of plasticity, *GDAM Report A11-S4*, Brown University, 1949.

10.10   A. Nadai, The principle of minimum work applied to states of finite, homo-geneous, plane plastic strain, *Proc. First U. S. Nat. Congr. Appl. Mech.*, 479–485, 1952.

10.11   P. Hodge and W. Prager, A variational principle for plastic materials with strain-hardening, *J. Math. and Phys.*, **27**, 1–10, 1948.

10.12   L. Finzi, Principle of minimum differential elastic energy, *Atti. Accad. Naz. Lincei R. C. Sci. Fis. Mat. Nat.* (8), **18**, 274–280, 1955. (Italian.)

11.1   W. T. Koiter, Stress-strain relations, uniqueness, and variational theorems for elastic-plastic materials with a singular yield surface, *Quart. Appl. Math.*, **11**, 350–354, 1953.

11.2   D. C. Drucker, On uniqueness in the theory of plasticity, *Quart. Appl. Math.*, **14**, 35–42, 1956.

11.3   R. Hill, On the problem of uniqueness in the theory of a rigid-plastic solid, *J. Mech. Phys. Solids*, 4, 247–255, 1956; 5, 1–8, 1956; 5, 153–161, 1957.

13.1   R. Hill, On the state of stress in a plastic-rigid body at the yield point, *Phil. Mag.* (7), **42**, 868–875, 1951.

13.2   R. Hill, A note on estimating yield-point loads in a plastic-rigid body, *Phil. Mag.* (7), **43**, 353–355, 1952.

13.3   H. J. Greenberg and W. Prager, Limit design of beams and frames, *Proc. Amer. Soc. Civil Engrs.*, **77**, Sep. No. 59, 1951.

13.4   D. C. Drucker, H. J. Greenberg, and W. Prager, The safety factor of an elastic-plastic body in plane strain, *J. Appl. Mech.*, **18**, 371–378, 1951.

13.5   D. C. Drucker, W. Prager, and H. J. Greenberg, Extended limit design theorems for continuous media, *Quart. Appl. Math.*, **9**, 381–389, 1952.

13.6   J. F. W. Bishop, On the complete solution to problems of deformation of a plastic-rigid material, *J. Mech. Phys. Solids*, **2**, 43–53, 1953.

13.7   E. H. Lee, On the significance of the limit load theorems for an elastic-plastic body, *Phil. Mag.* (7), **43**, 549–560, 1952.

13.8   D. C. Drucker, H. J. Greenberg, E. H. Lee, and W. Prager, On plastic-rigid solutions and limit design theorems for elastic-plastic bodies, *Proc. First U. S. Nat. Congr. Appl. Mech.*, 533–538, 1952.

13.9   T. Y. Thomas, Singular surfaces and flow lines in the theory of plasticity, *J. Rat. Mech. Anal.*, **2**, 339–381, 1953.

13.10 R. Hill, On the limits set by plastic yielding to the intensity of singularities of stress, *J. Mech. Phys. Solids.*, **2**, 278–285, 1954.

13.11 W. Prager, Discontinuous fields of plastic stress and flow, *Proc. Second U. S. Nat. Congr. Appl. Mech.*, 21–32, 1955.

13.12 J. L. Erickson, Singular surfaces in plasticity, *J. Math. and Phys.*, **34**, 74–79, 1955.

CHAP. 5.   BENDING OF A CIRCULAR PLATE

14.1   H. G. Hopkins and W. Prager, The load carrying capacities of circular plates, *J. Mech. Phys. Solids*, **2**, 1–13, 1953.

14.2   H. G. Hopkins and A. J. Wang, Load-carrying capacities for circular plates of perfectly-plastic material with arbitrary yield condition, *J. Mech. Phys. Solids*, **3**, 117–129, 1955.

14.3   D. C. Drucker and H. G. Hopkins, Combined concentrated and distributed load on ideally-plastic circular plates, *Proc. Second U. S. Nat. Congr. Appl. Mech.*, 517–520, 1955.

15.1   P. M. Naghdi, Bending of elastoplastic circular plates with large deflection, *J. Appl. Mech.*, **19**, 293–300, 1952.

15.2   H. G. Hopkins, Large elastic-plastic deformations of built-in circular plates under uniform load; Part I—theoretical analysis, *DAM Report DA-19-020-ORD-2598/12*, Brown University, 1954.

15.3   R. M. Haythornthwaite, The deflection of plates in the elastic-plastic range, *Proc. Second U. S. Nat. Congr. Appl. Mech.*, 521–526, 1955.

15.4   B. Tekinalp, Elastic, plastic bending of a simply supported circular plate under a uniformly distributed load, *DAM Report C11-6*, Brown University, 1955.

16.1   W. Prager, A new method of analyzing stresses and strains in work-hardening plastic solids, *J. Appl. Mech.*, **23**, 493–496, 1956.

16.2   W. Boyce, The bending of a work-hardening circular plate by a uniform transverse load, *Quart. Appl. Math.*, **14**, 277–288, 1956.

16.3   E. T. Onat and R. M. Haythornthwaite, Load carrying capacity of circular plates at large deflection, *J. Appl. Mech.*, **23**, 49–55, 1956.

17.1   H. G. Hopkins and W. Prager, On the dynamics of plastic circular plates, *Z. angew. Math. Phys.*, **5**, 317–330, 1954.

17.2   H. G. Hopkins, On the impact loading of circular plates made of a ductile material, *DAM Report DA-19-020-ORD-2598/7*, Brown University, 1954.

17.3   A. J. Wang and H. G. Hopkins, On the plastic deformation of built-in circular plates under impulsive load, *J. Mech. Phys. Solids*, **3**, 22–37, 1954.

18.1   H. Jung, Über eine Anwendung der Hillschen Minimalbedingung in der Plastizitätstheorie, *Ingen.-Arch.*, **23**, 61–68, 1955.

18.2   P. G. Hodge, Jr., and S. V. Nardo, Carrying capacity of an elastic-plastic cylindrical shell with linear strain hardening, *J. Appl. Mech.* (to be publ.)

CHAP. 6.   OTHER PROBLEMS
Shells

19.1   P. G. Hodge, Jr., The rigid-plastic analysis of symmetrically loaded cylindrical shells, *J. Appl. Mech.*, **21**, 336–342, 1954.

19.2   P. G. Hodge, Jr., Displacements in an elastic-plastic cylindrical shell, *J. Appl Mech.*, **23**, 73–79, 1956.

19.3   P. G. Hodge, Jr., and F. A. Romano, Deformations of an elastic-plastic cylindrical shell with linear strain hardening, *J. Mech. Phys. Solids*, **4**, 145–161, 1956.

19.4   P. G. Hodge, Jr., Impact pressure loading of rigid-plastic cylindrical shells, *J. Mech. Phys. Solids*, **3**, 176–188, 1955.

19.5   P. G. Hodge, Jr., Ultimate dynamic load of a circular cylindrical shell, *Proc. Second Midwest Conf. Solid Mech.*, 150–177, 1956.

19.6   P. G. Hodge, Jr., The influence of blast characteristics on the final deformation of circular cylindrical shells, *J. Appl. Mech.*, **23**, 617–624, 1956.

19.7   G. Eason and R. T. Shield, Dynamic loading of rigid-plastic cylindrical shells, *J. Mech. Phys. Solids*, **4**, 53–71, 1956.

19.8   D. C. Drucker, Limit analysis of cylindrical shells under axially-symmetric loading, *Proc. First Midwest Conf. Solid Mech.*, 158–163, 1953.

19.9   G. Eason and R. T. Shield, The influence of free ends on the load-carrying capacities of cylindrical shells, *J. Mech. Phys. Solids*, **4**, 17–27, 1955.

19.10   E. T. Onat, Plastic collapse of cylindrical shells under axially symmetrical loading, *Quart. Appl. Math.*, **13**, 63–72, 1955.

19.11   E. T. Onat and W. Prager, Limit analysis of shells of revolution, *Proc. Roy. Netherlands Acad. Sci.*, B, **57**, 534–548, 1954.

Plane strain and plane stress

20.1   W. Prager, A geometrical discussion of the slip line field in plane plastic flow, *Trans. Roy. Inst. Technol., Stockholm*, No. 65, 1953.

20.2   W. Prager, *Probleme der Plastitätstheorie*, Birkhauser Verlag, Basel and Stuttgart, 1955.

20.3   A. P. Green, On the use of hodographs in problems of plane plastic strain, *J. Mech. Phys. Solids*, **2**, 73–80, 1954.

20.4   H. Geiringer, Some recent results in the theory of an ideal plastic body, *Advances in Appl. Mech.*, III, 197–294, 1953.

20.5   E. H. Lee, Plastic flow in a V-notched bar pulled in tension, *J. Appl. Mech.*, **19**, 331–336, 1952.

20.6   E. H. Lee and A. J. Wang, Plastic flow in deeply notched bars with sharp internal angles, *Proc. Second U. S. Nat. Congr. Appl. Mech.*, 489–497, 1955.

20.7   L. Garr, E. H. Lee, and A. J. Wang, The pattern of plastic deformation in a deeply notched bar with semicircular roots, *J. Appl. Mech.*, **23**, 56–58, 1956.

20.8   R. Hill, On discontinuous plastic states, with special reference to localized necking in thin sheets, *J. Mech. Phys. Solids*, **1**, 19–30, 1952.

20.9   E. T. Onat and W. Prager, The necking of a tension specimen in plane plastic flow, *J. Appl. Phys.*, **25**, 491–493, 1954.

20.10   T. Y. Thomas, On the inclination of plastic slip bands in flat bars in tension tests, *Proc. Nat. Acad. Sci. U. S.*, **39**, 257–265, 1953.

20.11   A. P. Green, The plastic yielding of notched bars due to bending, *Quart. J. Mech. Appl. Math.*, **6**, 223–239, 1953.

20.12   A. P. Green, A theory of plastic yielding due to bending of cantilevers and beams, *J. Mech. Phys. Solids*, Part I, **3**, 1–15, 1954; Part II, **3**, 143–155, 1955.

20.13   F. A. Gaydon, An analysis of the plastic bending of a thin strip in its plane, *J. Mech. Phys. Solids*, **1**, 103–112, 1953.

20.14   J. W. Craggs, The influence of compressibility in elastic-plastic bending, *Quart. J. Mech. Appl. Math.*, **4**, 241–247, 1951.

20.15   M. R. Horne, The plastic bending of mild steel beams with particular reference to the effect of shear forces, *Proc. Roy. Soc. (London)*, A, **207**, 216–228, 1951.

20.16   E. T. Onat and R. T. Shield, The influence of shearing forces on the plastic bending of wide beams, *Proc. Second U. S. Nat. Congr. Appl. Mech.*, 535–537, 1955.

20.17   F. A. Gaydon, On the yield-point loading of a square plate with concentric circular hole, *J. Mech. Phys. Solids*, **2**, 170–176, 1954.

20.18   F. A. Gaydon and A. W. McCrum, A theoretical investigation of the yield point loading of a square plate with a central circular hole, *J. Mech. Phys. Solids*, **2**, 156–169, 1954.

20.19   P. G. Hodge, Jr., Final report on yield loads of slabs with reinforced cutouts, *GDAM Report B11-22*, Brown University, 1953.

20.20   H. J. Weiss, W. Prager, and P. G. Hodge, Jr., Limit design of a full reinforcement for a circular cutout in a uniform slab, *J. Appl. Mech.*, **19**, 397–401, 1952.

20.21   P. G. Hodge, Jr., and N. Perrone, Yield loads of slabs with reinforced cutouts, *J. Appl. Mech.*, **24**, 85–92, 1957.

20.22   P. G. Hodge, Jr., The effect of strain hardening in an annular slab, *J. Appl. Mech.*, **20**, 530–536, 1953.

20.23   E. H. Lee and B. W. Shaffer, The theory of plasticity applied to a problem of machining, *J. Appl. Mech.*, **18**, 405–413, 1951.

20.24   E. H. Lee, A plastic-flow problem arising in the theory of discontinuous machining, *Trans. ASME*, **76**, 189–193, 1954.

20.25   B. W. Shaffer, The mechanics of the simple shearing process during orthogonal machining, *Trans. ASME*, **77**, 331–336, 1955.

20.26   R. T. Shield and D. C. Drucker, The application of limit analysis to punch indentation problems, *J. Appl. Mech.*, **20**, 453–460, 1953.

20.27   R. T. Shield, The plastic indentation of a layer by a flat punch, *Quart. Appl. Math.*, **13**, 27–46, 1955.

20.28   E. W. Ross, Jr., On the ideally plastic indentation of inset rectangular bands, *J. Appl. Mech.*, **23**, 244–246, 1956.

20.29   E. Levin, Indentation pressure of a smooth circular punch, *Quart. Appl. Math.*, **13**, 133–137, 1955.

Beams, bars, and rods

21.1   P. S. Symonds and B. G. Neal, Recent progress in the plastic methods of structural analysis, *J. Franklin Inst.*, **252**, 383–407, 469–492, 1951.

21.2   N. J. Hoff, Complementary energy analysis of the failing load of a clamped beam, *J. Appl. Mech.*, **19**, 563–564, 1952.

21.3   E. T. Onat and W. Prager, The influence of axial forces on the collapse loads of frames, *Proc. First Midwest Conf. Solid Mech.*, 40–42, 1953.

21.4   A. W. Hendry, The plastic design of two-pinned mild steel arch ribs, *Civ. Eng. (London)*, **47**, 38–41, 1952.

21.5   E. T. Onat and W. Prager, Limit analysis of arches, *J. Mech. Phys. Solids*, **1**, 77–89, 1953.

21.6   C. Hwang, Plastic collapse of thin rings, *J. Aeronaut. Sci.*, **20**, 819–826, 1953.

21.7   W. Swida, Die elastisch-plastische Biegung des Krummen Stabes, *Ingen.-Arch.*, **16**, 357–372, 1948.

21.8   W. Swida, Die elastisch-plastische Biegung des Krummen Stabes unter Beruecksichtigung der Materialverfestigung, *Ingen.-Arch.*, **17**, 343–352, 1949.

21.9   W. Swida, Ueber die Restspannungen bei der elastisch-plastischen Biegung des Krummen Stabes, *Ingen.-Arch.*, **18**, 77–83, 1950.

21.10   I. Ohno, Stress calculation of a curved beam in a state of yielding, *Proc. First Jap. Nat. Congr. Appl. Mech.* (1951), 135–139, 1952.

21.11 A. Phillips, Bending with axial force of curved bars in plasticity, *J. Appl. Mech.*, **19**, 327–330, 1952.

21.12 B. W. Shaffer and R. N. House, Jr., The elastic-plastic stress distribution within a wide curved bar subjected to pure bending, *J. Appl. Mech.*, **22**, 305–310, 1955.

21.13 E. H. Lee and P. S. Symonds, Large plastic deformations of beams under transverse impact, *J. Appl. Mech.*, **19**, 308–314, 1952.

21.14 P. S. Symonds, Dynamic load characteristics in plastic bending of beams, *J. Appl. Mech.*, **20**, 475–481, 1953.

21.15 P. S. Symonds and C.-F. A. Leth, Impact of finite beams of ductile metal, *J. Mech. Phys. Solids*, **2**, 92–102, 1954.

21.16 J. A. Seiler and P. S. Symonds, Plastic deformation in beams under distributed dynamic loads, *J. Appl. Phys.*, **25**, 556–563, 1954.

21.17 M. F. Conroy, Plastic-rigid analysis of a special class of problems involving beams subject to dynamic transverse loading, *J. Appl. Mech.*, **22**, 48–52, 1955.

21.18 H. H. Bleich and M. G. Salvadori, Impulsive motion of elastoplastic beams, *Proc. Amer. Soc. Civil Engrs.*, **79**, Sep. No. 287, 1953.

21.19 E. W. Parkes, The permanent deformation of a cantilever struck transversely at its tip, *Proc. Roy. Soc. (London)*, A, **228**, 462–476, 1955.

21.20 T. J. Mentel, Plastic deformations due to dynamic loading of a beam with an attached mass, *Canad. J. Technol.*, **33**, 237–255, 1955.

21.21 E. T. Onat, Torsion of prismatic rods of work-hardening material, Thesis, Istanbul Technical University, 1951.

21.22 B. R. Seth, Finite elastic-plastic torsion, *J. Math. and Phys.*, **31**, 84–90, 1952.

21.23 R. Hill, The plastic torsion of anisotropic bars, *J. Mech. Phys. Solids*, **2**, 87–91 1954.

21.24 D. O. Brush, O. M. Sidebottom, and J. O. Smith, Plastic behavior of engineering materials. Part 1, Axial tension and bending interaction curves for members loaded inelastically, *Tech. Rep. 52-89*, Wright Air. Develop. Center, 1952.

21.25 A. J. Barrett, Unsymmetrical bending and bending combined with axial loading of a beam of rectangular cross section into the plastic range, *J. Roy. Aeronaut. Soc.*, **57**, 503–509, 1953.

21.26 J. M. Frankland and R. E. Roach, Strength under combined tension and bending in the plastic range, *J. Aeronaut. Sci.*, **21**, 449–453, 474, 1954.

21.27 H. Mii, Plastic deformation of light metal bars strained with combined tension and torsion, *J. Jap. Soc. Appl. Mech.*, **3**, 196–198, 204, 1950; **5**, 13–15, 1952.

21.28 F. A. Gaydon, On the combined torsion and tension of a partly plastic circular cylinder, *Quart. J. Mech. Appl. Math.*, **5**, 29–41, 1952.

21.29 R. Hill and M. P. L. Siebel, On the plastic distortion of solid bars by combined bending and twisting, *J. Mech. Phys. Solids*, **1**, 207–214, 1953.

21.30 H. Mii, Plastic deformation of light-metal bars strained with combined bending and torsion, *J. Jap. Soc. Appl. Mech.*, **5**, 11–14, 1952.

21.31 M. C. Steele, The plastic bending and twisting of square section members, *J. Mech. Phys. Solids*, **3**, 156–166, 1955.

21.32 J. Foulkes, Linear programming and structural design, *Proc. Second Symp. Linear Programming*, 177–184, 1955.

21.33 G. Kron, Solution of complex nonlinear plastic structures by the method of tearing, *J. Aeronaut. Sci.*, **23**, 557–562, 1956.

Tubes and cylinders

22.1　R. Hill and M. P. L. Siebel, On combined bending and twisting of thin tubes in the plastic range, *Phil. Mag.* (7), **42**, 722–733, 1951.

22.2　M. P. L. Siebel, The combined bending and twisting of thin cylinders in the plastic range, *J. Mech. Phys. Solids*, **1**, 189–206, 1953.

22.3　E. T. Onat and R. T. Shield, Remarks on combined bending and twisting of thin tubes in the plastic range, *J. Appl. Mech.*, **20**, 345–348, 1953.

22.4　B. Crossland and R. Hill, On the plastic behavior of thick tubes under combined torsion and internal pressure, *J. Mech. Phys. Solids*, **2**, 27–38, 1953.

ᵥ 22.5　D. G. Christopherson and G. R. Higginson, The strength of short cylinders under internal pressure, *J. Mech. Phys. Solids*, **2**, 217–237, 1954.

Axial symmetry

22.6　P. S. Symonds, On the general equations of problems of axial symmetry in the theory of plasticity, *Quart. Appl. Math.*, **6**, 448–452, 1949.

22.7　H. Jung, Axially symmetrical elastic-plastic body, *Österr. Ingen.-Arch.*, **7**, 168–180, 1953.

22.8　R. T. Shield, On the plastic flow of metals under conditions of axial symmetry, *Proc. Roy. Soc. (London)*, A, **233**, 267–287, 1955.

Metal forming

22.9　C. T. Yang and E. G. Thomsen, Plastic flow in a lead extrusion, *Trans. ASME*, **74**, 575–579, 1953.

22.10　E. G. Thomsen and J. Frisch, Stresses and strains in cold-extruding 2S-O aluminum, *Trans. ASME*, **77**, 1343–1353, 1955.

22.11　E. G. Thomsen, C. T. Yang, and J. B. Bierbower, An experimental investigation of the mechanics of plastic deformation of metals, *Univ. Calif. Publs. Eng.*, **5**, 89–144, 1954.

22.12　E. G. Thomsen, A new approach to metal forming problems, *Trans. ASME*, **77**, 515–522, 1955.

22.13　R. T. Shield, Plastic flow in a converging conical channel, *J. Mech. Phys. Solids*, **3**, 246–258, 1955.

22.14　H. W. Swift, Stresses and strains in tube drawing, *Phil. Mag.*, **40**, 883–902, 1949.

22.15　R. Hill, The calculation of stresses in the ironing of metal cups, *J. Iron Steel Inst.*, **161**, 41–44, 1949.

22.16　S. Y. Chung and H. W. Swift, Cup drawing from a flat bank, *Inst. Mech. Engrs. (London), Proc. Appl. Mech.*, **165**, 199–223, 1951.

22.17　Y. Yamada, Theory of formability testing of sheet metals, *Proc. Second Jap. Nat. Congr. Appl. Mech.* (1952), 51–56, 1953.

22.18　R. Hill, A theory of the plastic bulging of a metal diaphragm by lateral pressure, *Phil. Mag.* (7), **41**, 1133–1142, 1950.

22.19　E. W. Ross, Jr., and W. Prager, On the theory of the bulge test, *Quart. Appl. Math.*, **12**, 86–91, 1954.

22.20　N. A. Weil and N. M. Newmark, Large plastic deformations of circular membranes, *J. Appl. Mech.*, **22**, 533–538, 1955.

CHAP. 7. RUSSIAN CONTRIBUTIONS
(Unless otherwise stated all articles are written in Russian.)

Sec. 24. 1936–1949

General theory

24.1 N. M. Beliaev, Theory of plastic deformation, *Izvest. Akad. Nauk S.S.S.R. 1937*, 49–70, 1937.

24.2 S. M. Feinberg, The principle of limiting stress, *Prikl. Mat. Mekh.*, **12**, 63–68, 1948.

24.3 I. V. Goldenblat, On a method in the theory of elastic and plastic deformations, *Doklady*, **61**, 1001–1004, 1948.

24.4 I. V. Goldenblat, On the equations of equilibrium for a plastic medium, *Prikl. Mat. Mekh.*, **13**, 113–114, 1949.

24.5 I. V. Goldenblat, Some general laws of a process of elastic-plastic deformation, *Doklady*, **68**, 1005–1008, 1949.

24.6 H. Hencky, Development and present state of the theory of plasticity, *Prikl. Mat. Mekh.*, **4**, 31–36, 1940.

24.7 A. A. Ilyushin, Some problems on the theory of plastic deformations, *Prikl. Mat. Mekh.*, **7**, 245–272, 1943. (English summary.)

24.8 A. A. Ilyushin, Relation between the theory of St. Venant-Levy-Mises and the theory of small elastico-plastic deformations, *Prikl. Mat. Mekh.*, **9**, 207–218, 1945. (English summary.)

24.9 A. A. Ilyushin, Theory of small elastico-plastic deformations, *Prikl. Mat. Mekh.*, **10**, 347–356, 1946. (English summary.)

24.10 A. A. Ilyushin, On the theory of plasticity in case of simple loading of plastic bodies with strain-hardening, *Prikl. Mat. Mekh.*, **11**, 293–296, 1947. (English summary.)

24.11 A. J. Ishlinsky, Plane deformation when hardening takes place according to the linear law, *Prikl. Mat. Mekh.*, **5**, 57–70, 1941. (English summary.)

24.12 L. M. Kachanov, Plastic-elastic state of solids, *Prikl. Mat. Mekh.*, **5**, 431–438, 1941. (English summary.)

24.13 L. M. Kachanov, Variation principles for plastic-elastic solids, *Prikl. Mat. Mekh.*, **6**, 187–196, 1942. (English summary.)

24.14 L. M. Kachanov, On the stress-strain relation in the theory of plasticity, *Doklady*, **54**, 309–310, 1946.

24.15 S. Khristianovich, The plane problem of the mathematical theory of plasticity in the case where the surface tractions are given along a closed contour, *Mat. Sbornik*, **1**, 511–534, 1936.

24.16 L. S. Leibenson, *Elements of the Mathematical Theory of Plasticity*, Gostechisdat, Moscow and Leningrad, 1943.

24.17 E. V. Mahover, Certain problems of the theory of plasticity of anisotropic media, *Doklady*, **58**, 209–212, 1947.

24.18 A. A. Markov, Variation principles in the theory of plasticity, *Prikl. Mat. Mekh.*, **11**, 339–350, 1947. (English summary.)

24.19 S. G. Mikhlin, "The mathematical theory of plasticity," article in *Nekotorye novye voprosy mekhaniki sploshnoi sredy*, Izd. Akad. Nauk S.S.S.R., Moscow-Leningrad, 1938, pp. 157–216.

24.20 P. M. Riz, Large deformations and plasticity, *Prikl. Mat. Mekh.*, **12**, 211–212, 1948.

24.21   P. M. Riz, The theory of elasticity for large deformations exceeding the limit of proportionality, *Doklady*, **59**, 223–225, 1948.

24.22   W. W. Sokolovsky, The theory of plasticity—outline of work done, *Prikl. Mat. Mekh.*, **9**, 495–508, 1945. (English summary.) (English version published in *J. Appl. Mech.*, **13**, 1–10, 1946.)

24.23   W. W. Sokolovsky, *Theory of Plasticity*, Izd. Akad. Nauk S.S.S.R., Moscow-Leningrad, 1946, 306 pp. (2nd ed., 1950.)

24.24   W. W. Sokolovsky, On a form of representation of the components of stress in the theory of plasticity, *Doklady*, **61**, 223–225, 1948.

24.25   W. W. Sokolovsky, Some problems of the theory of plasticity for a strain-hardening material which obeys a power law, *Prikl. Mat. Mekh.*, **13**, 655–658, 1949.

Torsion

24.26   L. A. Galin, Elastico-plastic torsion of prismatic bars with polygonal cross-section, *Prikl. Mat. Mekh.*, **8**, 307–322, 1944. (English summary.)

24.27   L. A. Galin, On the existence of a solution of the elastic-plastic problem of torsion of prismatic bars, *Prikl. Mat. Mekh.*, **13**, 650–654, 1949.

24.28   L. M. Kachanov, The plastic torsion of circular rods of variable diameter, *Prikl. Mat. Mekh.*, **12**, 375–384, 1948.

24.29   B. A. Sokolov, Problems of elastic torsion of bars, *Prikl. Mat. Mekh.*, **8**, 468–474, 1944. (English summary.)

24.30   W. W. Sokolovsky, On the allowance for strain hardening of material in the problem of elastic-plastic torsion, *Doklady*, **36**, 46–51, 1942.

24.31   W. W. Sokolovsky, Ueber ein Problem der elastisch-plastischen Torsion, *Prikl. Mat. Mekh.*, **6**, 241–246, 1942. (German summary.)

24.32   W. W. Sokolovsky, Plastic torsion of a shaft of circular cross section and variable diameter, *Prikl. Mat. Mekh.*, **9**, 343–346, 1945. (English summary.)

Thick-walled spheres and cylinders

24.33   F. A. Bahsiyan, Finite displacements in a hollow sphere subjected to internal pressure, *Prikl. Mat. Mekh.*, **12**, 137–140, 1948.

24.34   N. M. Beliaev and A. K. Sinitsky, Stresses and strains in thick-walled cylinders in the elastic-plastic state, *Izvest. Akad. Nauk S.S.S.R. 1938*, **2**, 3–54, 1938.

24.35   N. M. Beliaev and A. K. Sinitsky, Stresses and strains in thick-walled cylinders in the elastic-plastic state with allowance for strain-hardening, *Izvest. Akad. Nauk S.S.S.R. 1938*, **4**, 21–49, 1938.

24.36   G. S. Shapiro, On the integration by quadratures of the equations of the plane one-dimensional problem of the theory of plasticity taking account of the hardening of the material, *Prikl. Mat. Mekh.*, **13**, 659–662, 1949.

24.37   W. W. Sokolovsky, Elastic-plastic state of the tube in the presence of strain-hardening in the material, *Doklady*, **37**, 160–165, 1942.

24.38   W. W. Sokolovsky, Elastic-plastic state of a cylindrical tube yielding with a strain-hardening of material, *Prikl. Mat. Mekh.*, **7**, 273–292, 1943. (English summary.)

24.39   W. W. Sokolovsky, The elastic-plastic equilibrium of a hollow sphere yielding the strain-hardening, *Prikl. Mat. Mekh.*, **8**, 70–78, 1944. (English summary.)

Rotating disks

24.40 Yu. N. Rabotnov, On a disk of equal resistance, *Prikl. Mat. Mekh.*, **12**, 463–464, 1948.

24.41 W. W. Sokolovsky, Plastic stresses in rotating disks, *Prikl. Mat. Mekh.*, **12**, 87–94, 1948.

Plane stress and plane strain

24.42 L. A. Galin, Plane elastico-plastic problem: Plastic zones in the vicinity of circular apertures, *Prikl. Mat. Mekh.*, **10**, 367–386, 1946. (English summary.)

24.43 L. A. Galin, An analogy for the plane elastic-plastic problem, *Prikl. Mat. Mekh.*, **12**, 757–760, 1948.

24.44 A. J. Ishlinsky, Plane strain in the case of linear strain-hardening, *Prikl. Mat. Mekh.*, **5**, 57–70, 1941.

24.45 O. S. Parasyuk, An elastic-plastic problem with a non-biharmonic plastic state, *Doklady*, **63**, 367–370, 1948.

24.46 K. N. Shevchenko, Approximate solution in closed form of a plane elastic-plastic problem with axial symmetry, *Prikl. Mat. Mekh.*, **13**, 323–328, 1949.

24.47 A. P. Sokolov, On an elastic-plastic state of a plate, *Doklady*, **60**, 33–36, 1948.

24.48 W. W. Sokolovsky, Plastic equilibrium equations of a plane stressed state, *Prikl. Mat. Mekh.*, **9**, 111–128, 1945. (English summary.)

24.49 W. W. Sokolovsky, Plastic equilibrium of a plane stressed state, *Prikl. Mat. Mekh.*, **10**, 209–220, 1946. (English summary.)

24.50 W. W. Sokolovsky, Equations of the plane plastic stressed state according to the Mises theory and their approximate representation, *Prikl. Mat. Mekh.*, **10**, 357–366, 1946. (English summary.)

24.51 W. W. Sokolovsky, Plastic plane stressed states according to Mises, *Doklady*, **51**, 175–178, 1946. (English.)

24.52 W. W. Sokolovsky, Plastic plane stressed state according to Saint-Venant, *Doklady*, **51**, 421–424, 1946. (English.)

24.53 W. W. Sokolovsky, A plane problem of the theory of plasticity on the distribution of stresses around a hole, *Prikl. Mat. Mekh.*, **13**, 159–164, 1949.

24.54 W. W. Sokolovsky, Approximate integration of the equations of a plane problem of the theory of plasticity, *Prikl. Mat. Mekh.*, **13**, 321–322, 1949.

Plates and shells

24.55 A. A. Ilyushin, Approximate theory of the elastico-plastic deformation of shells with the axial symmetry, *Prikl. Mat. Mekh.*, **8**, 15–24, 1944. (English summary.)

24.56 A. A. Ilyushin, Finite relationship between stresses and its relation with deformations in theory of shells, *Prikl. Mat. Mekh.*, **9**, 101–110, 1945. (English summary.)

24.57 V. M. Panferov, On the convergence of the method of elastic solutions in the theory of elastic-plastic deformations of shells, *Prikl. Mat. Mekh.*, **13**, 79–94, 1949.

24.58 W. W. Sokolovsky, Elastico-plastic bending of circular plates and plane rings, *Prikl. Mat. Mekh.*, **8**, 141–166, 1944. (English summary.)

24.59 W. W. Sokolovsky, On a problem of elastico-plastic bending of plates, *Doklady*, **52**, 13–16, 1946. (English.)

Rolling, drawing, and extruding processes

24.60 A. A. Ilyushin, Forming of tubes, *Inzhen. Sbornik*, **1**, 37–42, 1941.

24.61 A. J. Ishlinsky, Rolling and drawing at high speeds, *Prikl. Mat. Mekh.*, **7**, 226–230, 1943.

24.62 K. N. Shevchenko, Application of the theory of plasticity to the rolling of metals, *Prikl. Mat. Mekh.*, **5**, 439–452, 1941. (English summary.)

24.63 K. N. Shevchenko, On the distribution of stresses in a rolled bar, *Prikl. Mat. Mekh.*, **6**, 381–394, 1942. (English summary.)

24.64 K. N. Shevchenko, The pulling force during rolling, *Prikl. Mat. Mekh.*, **7**, 389–392, 1943.

24.65 K. N. Shevchenko, The plastic stressed state and the flow of metals in cold rolling and drawing, *Izvest. Akad. Nauk S.S.S.R. 1946*, 329–354, 1946.

Problems of contact stress and indentation

24.66 A. J. Ishlinsky, The problem of plasticity with the axial symmetry and Brinell's test, *Prikl. Mat. Mekh.*, **8**, 201–224, 1944. (English summary.)

24.67 K. N. Shevchenko, The elastic-plastic state due to a concentrated force applied to a half-plane, *Doklady*, **61**, 29–30, 1948.

24.68 K. N. Shevchenko, A concentrated force applied to a half plane (elastic-plastic problem), *Prikl. Mat. Mekh.*, **12**, 385–388, 1948.

24.69 W. W. Sokolovsky, Ueber den Druck eines plastischen Kontinuums auf einem harten Stempel, *Prikl. Mat. Mekh.*, **4**, 19–34, 1940. (German summary.)

Structural stability in the plastic range

24.70 A. A. Ilyushin, Stability of plates and shells stressed beyond the elastic limit, *Prikl. Mat. Mekh.*, **8**, 337–360, 1944. (English summary.)

24.71 A. A. Ilyushin, Elastico-plastic stability of plates, *Prikl. Mat. Mekh.*, **10**, 623–638, 1946. (English summary.)

24.72 I. P. Kuntze, Stability of plates of compressible materials beyond the limit of elasticity, *Prikl. Mat. Mekh.*, **10**, 671–672, 1946. (English summary.)

24.73 I. P. Kuntze, The stability of a cylindrical shell beyond the elastic limit, *Prikl. Mat. Mekh.*, **11**, 561–562, 1947.

24.74 I. P. Kuntze, Stabilité des plaques comprimées satisfaisant à la théorie de plasticité de Prager, *Doklady*, **55**, 387–389, 1947.

Sec. 25. 1949–1955

General theory

25.1 N. I. Bezuhov, *Theory of Elasticity and Plasticity*, Gosudarst. Izd. Tehn.-Teor. Lit., Moscow, 1953.

25.2 I. A. Birger, Certain general methods of solution of problems in the theory of plasticity, *Prikl. Mat. Mekh.*, **15**, 765–770, 1951.

25.3 M. I. Chernyak, An analytic expression of the volume strain under stretching in the elastic-plastic region, *Dopovidi Akad. Nauk Ukrain. R.S.R.*, 43–45, 1955. (Ukrainian, with Russian summary.)

25.4 N. S. Fastov, On the thermodynamics of plastic deformation, *Dokl. Akad. Nauk S.S.S.R. (N.S.)*, **78**, 251–254, 1951.

25.5 N. S. Fastov, On the equations of the theory of plasticity taking account of temperature variation, *Dokl. Akad. Nauk S.S.S.R. (N.S.)*, **85**, 67–70, 1952.

25.6   M. M. Filonenko-Borodich, On the conditions of the strength of materials possessing different tensile and compressive strengths, *Inzhen. Sbornik*, **19**, 13–36, 1954.

25.7   G. Ya. Galin, Surface conditions of strong disturbances for elastic and plastic bodies, *Prikl. Mat. Mekh.*, **19**, 368–370, 1955.

25.8   I. V. Goldenblat, The theory of small elastic-plastic deformations of anisotropic media, *Izvest. Akad. Nauk S.S.S.R. Otdel. Tehn. Nauk 1955*, 60–67, 1955.

25.9   I. V. Goldenblat, On the theory of small elastic-plastic deformations of anisotropic media, *Dokl. Akad. Nauk S.S.S.R.* (*N.S.*), **101**, 619–622, 1955.

25.10   S. I. Gubkin, Methods of determining deformability, *Izvest. Akad. Nauk S.S.S.R. Ser. Tekh. Nauk 1948*, 1463–1482, 1948.

25.11   S. I. Gubkin, Some basic problems of the science of plasticity, *Izvest. Akad. Nauk S.S.S.R. Otdel. Tehn. Nauk 1950*, 770–784, 1950.

25.12   S. I. Gubkin, Diagrams of schemes of mechanical states, *Izvest. Akad. Nauk S.S.S.R. Otdel. Tehn. Nauk 1950*, 1165–1182, 1950.

25.13   A. A. Ilyushin, The theory of elastic-plastic deformation and its applications, *Izvest. Akad. Nauk S.S.S.R. Otdel. Tehn. Nauk 1948*, 769–788, 1948.

25.14   A. A. Ilyushin, *Plasticity, Part One. Elastic-Plastic Deformations.* OGIZ, Moscow-Leningrad, 1948.

25.15   A. A. Ilyushin, Some fundamental problems of the theory of plasticity, *Izvest. Akad. Nauk S.S.S.R. Otdel. Tehn. Nauk 1949*, 1753–1773, 1949.

25.16   A. A. Ilyushin, Remarks on some papers devoted to a critique of the theory of plasticity, *Izvest. Akad. Nauk S.S.S.R. Otdel. Tehn. Nauk 1950*, 940–951, 1950.

25.17   A. A. Ilyushin, Modern problems in the theory of plasticity, *Vestnik Moscov. Univ.*, Nos. 4–5, 101–113, 1955. (See also *Sowjetwissenschaff. Naturwiss. Abt.*, 62–76, 1956, for a German translation.)

25.18   D. D. Ivlev, On the theory of simple deformation of plastic bodies, *Prikl. Mat. Mekh.*, **19**, 734–735, 1955.

25.19   W. M. Keldysch, editor, *Berechnung von Baukonstruktionen nach den Grenzbeanspruchungen* (translated from the Russian book *Limit Design of Structures* by W. A. Baldin, I. V. Goldenblat, W. M. Kotschenow, M. J. Pildisch, and K. E. Tal), VEB Verlag Technik, Berlin, 1953.

25.20   S. T. Kishkin and S. I. Ratner, Experimental check of the fundamental law of the plasticity theory, *Zhur. Tekh. Fiz.*, **19**, 412–420, 1949.

25.21   S. T. Kishkin, *Izvest. Akad. Nauk S.S.S.R. Odtel. Tehn. Nauk 1950*, 266–278, 1950.

25.22   D. Kuznecov, Concerning the lines of development of the theory of plasticity, *Izvest. Akad. Nauk S.S.S.R. Otdel. Tehn. Nauk 1950*, 760–769, 1950.

25.23   V. V. Moskvitin, On secondary plastic deformations, *Prikl. Mat. Mekh.*, **16**, 323–330, 1952.

25.24   V. V. Novozhilov, On the physical sense of the stress invariants used in the theory of plasticity, *Prikl. Mat. Mekh.*, **16**, 617–619, 1952.

25.25   V. V. Novozhilov, Class of strain paths characterized by the invariability of principal direction, *Prikl. Mat. Mekh.*, **18**, 415–424, 1954.

25.26   W. Olszak, Foundation of the theory of inhomogeneous elasto-plastic bodies, *Arch. Mech. Stos.* (Warsaw), **6**, 493–532, 1954 (Polish); *Bull. Acad. Polon. Sci.* (Warsaw), 111–117, 1955.

25.27   W. Olszak, On a classification of non-homogeneous elasto-plastic bodies, *Bull. Acad. Polon. Sci.* (Warsaw), **4**, 29–35, 1956. (Polish.)

25.28   W. Olszak, On the bases of the theory of non-homogenous elastic-plastic bodies, *Bull. Acad. Polon. Sci.* (Warsaw), 45–49, 1955. (Polish.)

25.29   V. M. Panferov, A general method of solution of boundary problems in the theory of elastic-plastic deformations for the simple loading of A. A. Ilyushin, *Vestnik Moskov. Univ.*, 41–62, 1952.

25.30   V. M. Panferov, On the applicability of variational methods to problems of the theory of small elastic-plastic deformations, *Prikl. Mat. Mekh.*, **16**, 319–322, 1952.

25.31   V. A. Pavlov and M. V. Yahutovich, Nature of "ductile" failure, *Dokl. Akad. Nauk S.S.S.R. (N.S.)*, **77**, 49–50, 1951.

25.32   P. P. Petrishchev, Elastic-plastic strains in an anisotropic body, *Vestnik Moskov. Univ.*, **7**, 63–72, 1952.

25.33   S. I. Ratner, *Izvest. Akad. Nauk S.S.S.R. Otdel. Tehn. Nauk 1950*, 435–450, 1950.

25.34   Yu. N. Rabotnov, Stress and deformation in cyclic loading, *Prikl. Mat. Mekh.*, **16**, 121–122, 1952.

25.35   K. V. Ruppeneit, On equations of the theory of plasticity for axisymmetrical problems, *Dokl. Akad. Nauk S.S.S.R. (N.S.)*, **80**, 557–560, 1951.

25.36   W. W. Sokolovsky, On the equations of the theory of plasticity, *Prikl. Mat. Mekh.*, **19**, 41–54, 1955.

25.37   I. I. Tarasenko, On the condition of failure of metals, *Zhur. Tekh. Fiz.*, **21**, 1336–1344, 1951.

25.38   S. D. Volkov, Generalized condition of plasticity, *Dokl. Akad. Nauk S.S.S.R. (N.S.)*, **79**, 213–216, 1951.

25.39   S. D. Volkov, On a condition for plasticity, *Dokl. Akad. Nauk S.S.S.R. (N.S.)*, **76**, 371–374, 1951.

Beams and rods

25.40   V. Ya. Bulygin, On elastic-plastic torsion of prismatic rods, *Prikl. Mat. Mekh.*, **16**, 107–110, 1952.

25.41   L. A. Galin, The elastoplastic torsion of prismatic bars, *Prikl. Mat. Mekh.*, **13**, 285–296, 1949.

25.42   A. I. Koshelev, Existence of a generalized solution for elastic and plastic torsional problems, *Dokl. Akad. Nauk S.S.S.R. (N.S.)*, **99**, 357–360, 1954.

25.43   B. S. Kovalskii, Elastoplastic deflection of a beam on an elastic foundation, *Dokl. Akad. Nauk S.S.S.R. (N.S.)*, **77**, 209–211, 1951.

25.44   V. V. Moskvitin, On elastic-plastic bending of a beam, *Vestnik Moskov. Univ. Ser. Fiz.-Mat. Estestven. Nauk*, **9**, 33–40, 1954.

25.45   Yu. I. Yagn and E. N. Tarasenko, Applied theory of plastic deformation of beams, *Dokl. Akad. Nauk S.S.S.R. (N.S.)*, **73**, 471–474, 1950.

25.46   M. Zyczkowski, The problem of combined tension and torsion of a circular bar in the elasto-plastic range, *Rozprawy Inzynierskie* (Warsaw), 285–322, 1955. (Polish with English and Russian summaries.)

25.47   M. Zyczkowski, Simultaneous tension and torsion of a circular bar in the elasto-plastic range, *Bull. Acad. Polon. Sci.* (Warsaw), 51–55, 1955. (English.)

Thick-walled spheres and cylinders

25.48   E. H. Agababyan, Stresses in a tube under a sudden application of a load, *Ukrain. Mat. Zhur.*, **5**, 325–332, 1953.

25.49   E. H. Agababyan, Dynamic extension of a hollow cylinder under conditions of ideal plasticity, *Ukrain. Mat. Zhur.*, **7**, 243–252, 1955.

25.50   L. M. Kachanov, On a problem of deformation in plastic layers, *Dokl. Akad. Nauk S.S.S.R. (N.S.)*, **96**, 249–252, 1954.

25.51 V. A. Lomakin, Large strains of a tube and of a hollow sphere, *Inzhen. Sbornik*, **21**, 61–73, 1955.

25.52 V. A. Lomakin, On large elastic-plastic deformations, *Vestnik Moskov. Univ. Ser. Fiz.-Mat. Estestven. Nauk*, **9**, 41–45, 1954.

25.53 Viktor Lovass-Nagy, Sur les états de tension plastiques et élastiques dans les tuyaux à paroi épaisse, *Magyar Tud. Akad. Alkalm. Mat. Int. Közl.*, **1**, 49–80, 1953. (Hungarian, with Russian and French summaries.)

25.54 W. Olszak and W. Urbanowski, A heterogeneous thick-walled elastic-plastic cylinder subjected to internal pressure and longitudinal force, *Arch. Mech. Stos.* (Warsaw), **7**, 315–336, 1955. (Polish, with summaries in English and Russian.)

25.55 W. Urbanowski, Elastic-plastic deformation of a thick-walled spherical vessel subjected to internal pressure, *Arch. Mech. Stos.* (Warsaw), **7**, 519–532, 1955. (Polish.)

Rotating disks

25.56 A. G. Kostyuk, Stresses in a continuous rotating cylinder beyond the elastic limit, *Prikl. Mat. Mekh.*, **18**, 453–456, 1954.

25.57 M. Sh. Mikeladze, Elastic and plastic deformations in fast rotating disks of varying thickness, *Inzhen. Sbornik*, **15**, 21–34, 1953.

25.58 M. Sh. Mikeladze, On the plastic state of a rotating anisotropic cylinder, *Prikl. Mat. Mekh.*, **19**, 504–506, 1955.

Plane stress and plane strain

25.59 A. M. Kochetkov, Stresses in a wedge under the influence of hydrostatic pressure, *Inzhen. Sbornik*, **15**, 177–180, 1953.

25.60 W. Olszak, The plane problem of the theory of plastic flow of nonhomogeneous bodies, *Bull. Acad. Polon. Sci.* (Warsaw), **3**, 119–124, 1955. (English.)

25.61 V. M. Panferov, The plane problem of the theory of small elastic-plastic deformations, *Acad. Repub. Pop. Romine. An. Romino-Soviet Mat.-Fiz.* (3), **7**, 29–53, 1954. (Romanian.) (Translated from *Vestnik Moskov. Univ. Ser. Fiz.-Mat. Estestven. Nauk 1953*, 45–68, 1953.)

25.62 V. M. Panferov, Concentration of stresses in elasto-plastic deformations, *Izvest. Akad. Nauk S.S.S.R. Otdel. Tekh. Nauk*, **4**, 47–66, 1954.

25.63 G. N. Savin and O. S. Parasyuk, On some elastic-plastic problems with linear hardening, *Doklady*, **70**, 585–588, 1950.

25.64 G. N. Savin and O. S. Parasyuk, Some elastic-plastic problems with linear hardening, *Ukrain. Mat. Zhurn.*, **2**, 60–69, 1950.

25.65 G. S. Shapiro, On the integration by quadratures of the equations of the plane one-dimensional problem of the theory of plasticity taking account of the hardening of the material, *Prikl. Mat. Mekh.*, **13**, 659–662, 1949.

25.66 G. S. Shapiro, Elastic-plastic equilibrium of a wedge and discontinuous solutions in the theory of plasticity, *Prikl. Mat. Mekh.*, **16**, 101–106, 1952.

25.67 W. W. Sokolovsky, Equations of plastic equilibrium for plane stress, *Prikl. Mat. Mekh.*, **13**, 219–221, 1949.

25.68 W. W. Sokolovsky, On a plane problem of the theory of plasticity, *Prikl. Mat. Mekh.*, **13**, 391–400, 1949.

25.69 W. W. Sokolovsky, Plane and axisymmetric equilibrium of a plastic mass between rigid walls, *Prikl. Mat. Mekh.*, **14**, 75–92, 1950.

25.70 W. W. Sokolovsky, Plane equilibrium of a plastic wedge, *Prikl. Mat. Mekh.*, **14**, 391–404, 1950.

25.71 W. W. Sokolovsky, Some remarks concerning a plane problem in the theory of plasticity, *Prikl. Mat. Mekh.*, **18**, 762–763, 1954.

Plates and shells

25.72 A. N. Ananina, Axially symmetric deformations of a cylindrical shell with elastic-plastic deformations, *Inzhen. Sbornik*, **18**, 157–160, 1954.

25.73 A. S. Grigorev, On the bending of a round elastic plate beyond the elastic limit, *Prikl. Mat. Mekh.*, **16**, 111–115, 1952.

25.74 A. S. Grigorev, Load-carrying capacity of thick flat plastic rings, *Inzhen. Sbornik*, **16**, 177–182, 1953.

25.75 A. S. Grigorev, Bending of circular and annular plates of varying thickness beyond the limits of elasticity, *Inzhen. Sbornik*, **20**, 59–92, 1954.

25.76 A. A. Ilyushin, Normal and tangential stress in the pure bending of beams beyond the limit of resiliency, and the analogy of the bending plates, *Inzhen. Sbornik*, **19**, 3–12, 1954.

25.77 B. G. Korenev, On the computation of beams and plates, taking account of plastic deformation, *Inzhen. Sbornik*, **5**, 58–61, 1948.

25.78 A. G. Kostyuk, On the equilibrium of an annular plate with a power law of hardening, *Prikl. Mat. Mekh.*, **14**, 319–320, 1950.

25.79 R. A. Mezhlumyan, Flexure and torsion of thin-walled cylindrical shells beyond the elastic limit, *Prikl. Mat. Mekh.*, **14**, 253–264, 1950.

25.80 R. A. Mezhlumyan, The boundary conditions in bending and torsion of thin shells beyond the elastic limit, *Prikl. Mat. Mekh.*, **14**, 537–542, 1950.

25.81 R. A. Mezhlumyan, Determination of the bearing capacity of thin-walled structures taking into account the material-hardening, *Prikl. Mat. Mekh.*, **15**, 175–182, 1951.

25.82 W. Olszak and A. Sawczuk, The theory of limit loads of plates, with reference to its experimental verification, *Rozprawy Inzynierskie* (Warsaw), 179–253, 1955. (Polish with French and Russian summaries.)

25.83 W. Olszak and A. Sawczuk, Experimental verification of the limit analysis of plates, *Bull. Acad. Polon. Sci.* (Warsaw), 195–200, 1955. (English.)

25.84 V. M. Panferov, Convergence of a method of solution for the elastoplastic bending of plates, *Prikl. Mat. Mekh.*, **16**, 195–212, 1952.

25.85 Yu. N. Rabotnov, Approximate technical theory of elastoplastic shells, *Prikl. Mat. Mekh.*, **15**, 167–174, 1951.

25.86 V. I. Rosenblyum, Approximate theory of equilibrium of plastic shells, *Prikl. Mat. Mekh.*, **18**, 289–302, 1954.

25.87 A. Sawczuk, Some problems of limit analysis of elements subjected to tension and bending and their application to the theory of rectangular tanks, *Rozprawy Inzynierskie* (Warsaw), 255–284, 1955. (Polish, with English and Russian summaries.)

25.88 K. N. Shevchenko, Axisymmetrical elastoplastic problem for a plate weakened by a circular hole, *Prikl. Mat. Mekh.*, **15**, 519–520, 1951.

Other problems

25.89 E. Bölcskei, Limit design of compressed bars, *Acta Tech. Acad. Sci. Hung.*, **15**, 19–35, 1956. (French and German summaries.)

25.90 Wei-Zang Chien and Chih-Ta Chen, Theory of rolling, *Chinese J. Phys.*, **9**, 57–92, 1953. (Chinese, English summary.)

25.91 S. I. Gubkin, Theory of flow stresses in metals at drawing, *Izvest. Akad. Nauk S.S.S.R. Ser. Tekh. Nauk 1947*, 1663–1681, 1947.

25.92 Yu. R. Lepik, Once again on the cylindrical form of buckling of elastic-plastic plates, *Prikl. Mat. Mekh.*, **20**, 140–143, 1956.

25.93 K. N. Shevchenko, Elasto-plastic problem for a heavy half-space with a vertical cylindrical opening, *Prikl. Mat. Mekh.*, **14**, 587–592, 1950.

25.94 K. N. Shevchenko, Plane elastic-plastic deformation of a cylinder loaded by a balanced system of two concentrated forces, *Prikl. Mat. Mekh.*, **16**, 35–44, 1952.

25.95 G. P. Slepcova, Plastic deformation of a circular membrane under static loading, *Dopovidi Akad. Nauk Ukrain. R.S.R.*, 520–524, 1955. (Ukrainian, with Russian summary.)

25.96 V. H. Sobolev and L. D. Sokolov, On the pressure of a rigid die on a plastic medium, *Inzhen. Sbornik*, **5**, 21–24, 1949.

25.97 K. V. Ruppeneit, Compression of a cylinder between two rough rigid plates, *Dokl. Akad. Nauk S.S.S.R. (N.S.)*, **72**, 247–250, 1950.

Author Index

Agababyan, E. H., 43, 141
Ananina, A. N., 143

Bahsiyan, F. A., 137
Baker, J. F., 123
Baldin, W. A., 140
Barrett, A. J., 120, 134
Barton, M. V., 21
Batdorf, S. B., 73, 129
Beliaev, N. M., 136, 137
Besseling, J. F., 73, 79, 129
Bezuhov, N. I., 139
Bierbower, J. B., 135
Birger, I. A., 139
Bishop, J. F. W., 91, 128, 130
Blake, F. G., 43
Bleich, H. H., 119, 123, 134
Bogdanoff, J. L., 26
Bohnenblust, H. F., 128
Bölcskei, E., 143
Born, J. S., 20
Boussinesq, J., 31
Boyce, W., 102, 131
Brush, D. O., 120, 134
Budiansky, B., 73, 129
Bulygin, V. Ya., 127, 141
Burmistrov, E. F., 5
Bycroft, G. N., 43

Chaplygin, S. A., 127
Chen, Chih-Ta, 143
Chernyak, M. I., 139
Chien, Wei-Zang, 143

Christie, D. G., 42
Christopherson, D. G., 120, 135
Chung, S. Y., 120, 135
Cicala, P., 129
Colonnetti, G., 81, 130
Conroy, M. F., 119, 134
Conway, H. D., 26
Craggs, J. W., 118, 132
Crossland, B., 120, 135

Danilovskaya, V. I., 28
Davies, R. M., 43
Dow, N. F., 129
Drucker, D. C., 59, 75, 87, 91, 118, 128, 129, 130, 131, 132, 133
Duwez, P., 128

Eason, G., 132
Edelman, F., 62, 128
Elliott, H. A., 25
Erickson, J. L., 131
Eshelby, J. D., 39
Eubanks, R. A., 25

Fastov, N. S., 139
Feinberg, S. M., 136
Filonenko-Borodich, M. M., 140
Finzi, L., 81, 130
Foulkes, J., 120, 134
Frankland, J. M., 120, 134
Freudenthal, A. M., 28, 128
Fridman, M. M., 21, 23, 41, 42
Frisch, J., 135

Galerkin, B. G., 127
Galin, G. Ya., 140
Galin, L. A., 3, 20, 24, 30, 31, 32, 33, 34, 35, 38, 122, 124, 125, 127, 137, 138, 141
Garr, L., 132
Gatewood, B. E., 27
Gaydon, F. A., 118, 120, 132, 133, 134
Geiringer, H., 118, 124, 132
Goldenblat, I. V., 136, 140
Goodier, J. N., 6, 7, 14, 21, 26, 42
Green, A. E., 3, 6, 11, 17, 22, 23, 24, 31, 44
Green, A. P., 118, 132
Greenberg, H. J., 81, 91, 130
Greenspan, M., 6
Grigorev, A. S., 143
Gubkin, S. I., 140, 143
Gutin, L. Ya., 43

Handelman, G. H., 62, 126, 128
Harding, J. W., 31
Harvey, R. B., 11
Haythornthwaite, R. M., 98, 131
Heller, S. R., 11
Hencky, H., 136
Hendry, A. W., 119, 133
Hertz, H., 19, 30
Hieke, M., 27
Higginson, G. R., 120, 135
Hill, R., 61, 81, 87, 91, 118, 119, 120, 124, 125, 128, 130, 131, 132, 134, 135
Hobson, E. W., 30, 35
Hodge, P. G., Jr., 75, 81, 118, 128, 129, 130, 131, 132, 133
Hoff, N. J., 119, 133
Horne, M. R., 118, 132
Hopkins, H. G., 18, 95, 98, 106, 131
Horvay, G., 20
House, R. N., Jr., 119, 134
Howland, R. C. J., 6
Hsu, C. S., 14
Huth, J. H., 27
Hwang, C., 119, 133

Ilyushin, A. A., 121, 126, 136, 138, 139, 140, 143
Ishlinsky, A. J., 136, 138, 139
Isida, M., 7
Ivlev, D. D., 140

Jahsman, W. E., 42
Jung, H., 109, 120, 131, 135

Kachanov, L. M., 81, 136, 137, 141
Kaechele, L., 129
Karman, Th. von, 125
Keldysch, W. M., 140
Khristianovich, S., 136
Kikukawa, M., 8, 10, 12, 13
Kishkin, S. T., 140
Kist, N. C., 123
Kitover, K. A., 20
Kochetkov, A. M., 142
Koiter, W. T., 14, 75, 85, 129, 130
Kolosov, G., 26
Korenev, B. G., 143
Koshelev, A. I., 141
Kosmodamianski, A. S., 23
Kostyuk, A. G., 142, 143
Kotschenow, W. M., 140
Kovalskii, B. S., 141
Kromm, A., 42
Kron, G., 120, 134
Kufarev, P. P., 24
Kuntze, I. P., 126, 139
Kusukawa, K., 38
Kuznecov, D., 140

Lamb, H., 42, 43
Lapwood, R. E., 43
Lebedev, N. N., 26
Lee, E. H., 91, 118, 119, 130, 132, 133, 134
Leibenson, L. S., 24, 136
Lekhnitzki, S. G., 3, 21, 23, 24, 25
Lepik, Yu. R., 144
Leth, C-F. A., 134
Levin, E., 119, 133
Ling, C. B., 6, 7
Lodge, A. S., 25
Lomakin, V. A., 142
Lovass-Nagy, Viktor, 142
Love, A. E. H., 30
Lourie, A. I., 6, 21, 33
Lubkin, J. L., 32

McCrum, A. W., 133
Mahover, E. V., 136
Maisel, W. M., 27
Mansfield, E. H., 14

Markov, A. A., 81, 136
Maue, A. W., 39, 40, 41
Melan, E., 26, 28, 123
Mentel, T. J., 119, 134
Mezhlumyan, R. A., 143
Mii, H., 120, 134
Mikeladze, M. Sh., 142
Mikhlin, S. G., 24, 136
Miller, G. F., 43
Mindlin, R. D., 6, 30
Mises, R. von, 61, 124, 127, 128
Morduchov, M., 11
Morkovin, D., 6
Moskvitin, V. V., 140, 141
Mossakowskii, V. I., 18, 34, 35
Mott, N. F., 39
Mura, T., 28
Muskhelishvili, N. I., 3, 4, 5, 7, 11, 14, 15, 17, 18, 20, 23, 24, 27, 125

Nabarro, F. R. N., 39
Nadai, A., 51, 81, 124, 128, 130
Naghdi, P. M., 98, 131
Nardo, S. V., 131
Narishkina, E., 43
Neal, B. G., 119, 133
Neuber, H., 4
Newlands, M., 43
Newmark, N. M., 120, 135
Novozhilov, V. V., 140

Odqvist, F. K. G., 124
Ohno, I., 119, 133
Olszak, W., 140, 142, 143
Onat, E. T., 118, 119, 120, 131, 132, 133, 134, 135
Owens, A. J., 23

Panferov, V. M., 138, 141, 142, 143
Papkovicz, P. F., 20
Parasyuk, O. S., 125, 138, 142
Parkes, E. W., 119, 134
Parkus, H., 26
Pavlov, V. A., 141
Payne, L. E., 25, 35
Perrone, N., 133
Peters, R. W., 129
Petrashen, G., 43
Petrishchev, P. P., 141
Philippidis, A. H. (See Phillips, A.)

Phillips, A., 75, 81, 119, 129, 130, 134
Pildisch, M. J., 140
Plemelj, J., 17
Prager, W., 55, 65, 75, 81, 91, 95, 102, 106, 118, 119, 120, 122, 123, 124, 125, 126, 128, 129, 130, 131, 132, 133, 135, 139
Pursey, H., 43

Rabotnov, Yu. N., 138, 141, 143
Radok, J. R. M., 3, 4, 7, 11, 13, 14, 15
Rankin, A. W., 21
Ratner, S. I., 140, 141
Reissner, E., 32, 43
Reissner, H., 11
Riz, P. M., 136, 137
Rivlin, R. S., 44
Roach, R. E., 120, 134
Roderick, J. W., 123
Romano, F. A., 132
Rosenblyum, V. I., 143
Ross, E. W., Jr., 119, 120, 133, 135
Ruppeneit, K. V., 141, 144

Sadowsky, M. A., 28, 81, 124, 130
Saenz, A. W., 39
Sagoci, H., 32
Salvadori, M. G., 119, 134
Sanders, J. L., Jr., 129
Sauter, F., 39
Savin, G. N., 3, 4, 5, 6, 7, 11, 12, 21, 23, 24, 142
Sawczuk, A., 143
Seiler, J. A., 134
Selberg, H. L., 43
Seremetev, M. P., 11, 13, 18
Seth, B. R., 119, 134
Shaffer, B. W., 118, 119, 133, 134
Shapiro, G. S., 21, 137, 142
Shepherd, R. P., 129
Sherman, D. I., 7, 27, 43
Shevchenko, K. N., 138, 139, 143, 144
Shield, R. T., 25, 35, 118, 119, 120, 132, 133, 135
Shtaerman, E., 20
Sidebottom, O. M., 120, 134
Siebel, M. P. L., 120, 134, 135
Sinitsky, A. K., 137
Slepcova, G. P., 144
Smith, C. B., 23

Smith, J. O., 120, 134
Smirnov, V. I., 41, 43
Sneddon, I. N., 18, 31, 37
Sobolev, S. L., 41, 43
Sobolev, V. H., 144
Sokolnikoff, I. S., 3, 24
Sokolov, A. P., 138
Sokolov, B. A., 137
Sokolov, L. D., 144
Sokolovsky, W. W., 52, 124, 127, 137, 138, 139, 141, 142, 143
Steele, M. C., 120, 134
Sternberg, E., 25
Stockton, F. D., 75, 129
Sveklo, V. A., 42
Swida, W., 119, 133
Swift, H. W., 120, 135
Symonds, P. S., 119, 120, 133, 134, 135

Tal, K. E., 140
Tarabasov, N. D., 27
Tarasenko, E. N., 141
Tarasenko, I. I., 141
Taylor, G. I., 22
Tekinalp, B., 98, 131
Thomas, T. Y., 61, 118, 128, 131, 132
Thomsen, E. G., 120, 135
Timoshenko, S., 6, 7, 21, 26, 56
Tresca, H., 96
Truesdell, C., 43
Tupper, S. J., 125

Udoguchi, T., 27
Ugodchikov, A. G., 27
Urbanowski, W., 142

Van den Broek, J. A., 123
Volkov, S. D., 141

Wang, A. J., 106, 131, 132
Warburton, G. B., 43
Warner, W. H., 62, 128
Weil, N. A., 120, 135
Weiner, J. H., 28
Weiss, H. J., 133
Wells, A. A., 11
White, G. N., Jr., 79, 129
Williams, M. L., 20

Yagn, Yu. I., 141
Yakutovich, M. B., 141
Yamada, Y., 120, 135
Yang, C. T., 135
Young, D. M., 8
Yu, Yi-Yuan, 7

Zagubizenko, P. A., 18
Zarantonello, H., 8
Zerna, W., 3, 11, 17, 22, 23, 24, 31, 44
Zvolinski, N. V., 38
Zyczkowski, M., 141

Subject index

Adhesion, 18, 20, 27, 30, 35
Admissible fields, *see* Geometrically, Kinematically, and Statically Admissible Fields
Anisotropic material, 21, 24
  medium, 5, 38
  plate, 13, 21
Anisotropy, 73, 119
Annulus, circular, 7
Axially symmetric problems, 56, 120, 135
  See also Circular cylindrical shell, Circular plate

Bars, 119, 120, 133, 141
Beams, 69, 118, 119, 133, 141
  combined stresses in, 69, 119
Bipolar co-ordinates, 7
Boundary conditions, mixed, 3, 14, 18, 38, 42
Boundary value problem, 62, 81, 85
Buckling, 51, 125, 139, 144
Bulging, 120, 135

Cavity, elliptic, 24
  spherical, 43
Chip formation, 118
Circular cylindrical shell, 109, 113, 124, 127, 131, 138, 143
Circular plate, 56, 95, 98, 102, 106, 109, 124, 131, 138, 143
Collapse load, 97, 100, 106, 118
Combined stresses, in beams, 69, 119
Complementary energy, 84, 90, 91

Complex potentials, 5
Cone, 25
Conformal mapping, 5, 8
Contact, 18, 20, 26, 27, 29, 30, 31, 32, 34
  arcs of, 18
Continuity requirements, 94, 100, 102, 104
Crack, elliptical, 25
Creep, 51
Curvilinear co-ordinates, 5
  bipolar, 27
Cuts, 24, 40
Cutouts, 118
Cylinder, 24, 25, 124, 127, 137
  circular, 21, 27, 119
  parabolic, 24
  thick-walled, 43
Cylindrical shell, 6

Deformation, finite, 44
Deformation theory, 61, 121, 124, 126, 127, 136, 139
Design, for minimum weight, 120
Die, rigid, 30
  traveling, 20
  See also Punch, Stamp
Diffraction, 41
Discontinuity, 94, 98
Disk, 18
  circular, 15
  heavy, 7
  oversize, 18, 27
  rotating, 138, 142

Disk, rotating elliptic, 23
with hot sector, 27
Dislocation, 27, 39
Dissipation of energy, 86, 87, 88
Drawing, 118, 120, 125, 135, 139, 143
Dynamic loading, 106, 119, 127

Eigenfunctions, 20
Eigensolutions, 21
Elastic-plastic material, 57, 77, 85, 98, 101
Elastic range, 64, 65
Elasticity, 52, 56, 59, 81
Energy, 59
    See also Complementary energy, Internal energy, Potential energy
Energy rate, 86, 87, 88
Expansion, coefficient of, 29
Extremum principles, see Minimum principles
Extrusion, 118, 120, 125, 135, 139

Fatigue, 28
Fillets, 8
Flexure, 25
Flow law, 60, 61, 63, 70, 72, 75, 121, 127, 136, 139
Fluid layer, 38
Force, traveling concentrated, 37
Forming processes, see Drawing, Extrusion, Rolling
Foundation, elastic, 43
Frames, 119
Friction, 18, 20, 24, 30, 32, 35, 38

Generalized strain, see Generalized variables
Generalized stress, see Generalized variables
Generalized variables, 55, 66, 69, 82, 95, 113
Geometrically admissible field, 82
Gravitational stress, 7

Half-plane, 23, 38
orthotropic, 24
Hardening, anisotropic, 73
isotropic, 64, 71, 77, 110
kinematic, 65, 69, 78, 79, 105
Heat conduction, sources, moving, 28
steady, 27

Hilbert problem, 17, 20, 22, 23, 24, 41, 42
Holes, 5, 8
circular, 5, 7, 13, 18, 23, 26, 27, 42
elliptical, 5, 7, 18, 23, 26
ovaloid, 6, 7, 8, 24
rectangular, 5
reinforcement of, 11
triangular, 5

Impact, 30
Inclusion, elliptic, 23, 24
Incompressibility, 84
Indentation, 118, 139
Internal energy, 55
Irreversibility, 59
Isotropic hardening, 64, 71, 77, 110
Isotropy, transverse, 25

Kinematic hardening, 65, 69, 78, 79, 105
Kinematic model, see Kinematic hardening
Kinematically admissible field, 85, 93, 94, 117

Lamé functions, 33
Laplace transformation, 28
Limit analysis, 91, 115, 117, 123, 126
Linear programming, 120
Loading, dynamic, 106, 119, 127
proportional, 67
Loading path, 61

Material, elastic-plastic, 57, 77, 85, 98, 101
perfectly plastic, 52, 57, 71, 75, 85, 89, 91, 95, 98, 106, 117, 124, 127, 128
rigid-plastic, 52, 57, 78
strain-hardening, 52, 63, 65, 77, 88, 91, 102, 110, 127, 129
Metal forming, see Drawing, Extrusion, Rolling
Minimum principles, 81, 86, 87, 88, 89, 91, 124, 127, 130
    See also Complementary energy, Potential energy
Minimum weight design, 120
Mises yield condition, 127
Multiply-connected regions, 7

Necking, 118
Notch, 8
  hyperbolic, 23

Orthotropy, cylindrical, 23

Perfectly plastic material, 52, 57, 71, 75, 85, 89, 91, 95, 98, 106, 117, 124, 127, 128
Piecewise linear plasticity, 65, 75, 117, 129
Plane strain, 4, 117, 118, 119, 122, 124, 125, 127, 132, 136, 138, 142
Plane stress, 4, 118, 119, 125, 127, 132, 138, 142
Plastic buckling, 51, 125, 139, 144
Plastic irreversibility, 59
Plastic strain rate, see Strain rate
Plastic waves, 51
Plasticity, piecewise linear, 65, 75, 117, 129
Plate, 35
  circular, 56, 95, 98, 102, 106, 109, 124, 131, 138, 143
  flexure, 5, 13, 21
  rectangular, 21
  semi-infinite, 18
Poisson's ratio, 56
Potential, displacement, 27
  logarithmic, 27
  Newtonian, 27, 29
Potential energy, 83, 90, 91, 109
Pressure, 30, 31, 38, 43
  band, 21
  surface, 25
Principle, of minimum energy, see Minimum principles
  of virtual work, 82
Problem of linear relationship, 17
Progression, regular, 76, 78, 89
Pulse, 40, 42
Punch, elliptical, 25
  indentation by, 118
  rigid, 30
  See also Die, Stamp

Reciprocal theorem, 27
Regular progression, 76, 78, 89
Riemann-Hilbert problem, 20, 24, 38
Rigid plastic material, 52, 57, 78

Ring, partial, 23
  reinforcing, 11, 13
Rolling, 118, 125, 127, 139, 143
Rotating disk, 138, 142
Rotationally symmetric problems, 56, 120, 135
  See also Circular cylindrical shell, Circular plate
Russian contributions, 121

Safety factor, 92, 123
Shell, axially symmetric, 56
  circular cylindrical, 109, 113, 124, 127, 131, 138, 143
Shock, thermal, 28
Shrink fit, eccentric, 27
Similarity, 39
Singular integral equations, 14, 17
Singular yield condition, 60, 72, 76, 85
Slab, 23, 25, 43
Slip, 30
Slip theory, 73
Slit, 39, 41
  circular arc, 18
Soil mechanics, 51
Sphere, 26
  thick-walled, 137, 141
Stamp, 31, 33, 35
  arbitrary plan form, 34
  elliptic, 34
  revolving, 32
  rigid, 18, 30, 38
  sectorial plan form, 35
  See also Die, Punch
Statically admissible field, 82, 85, 92, 115
Strain, plane, 117, 118, 119, 122, 125, 127, 132, 136, 138, 142
  plastic, 61, 65, 67, 71, 76, 78
Strain energy, 26
Strain hardening, see Hardening
Strain-hardening material, 52, 63, 65, 77, 88, 91, 102, 110, 127, 129
Strain rate, 57, 58, 63, 72, 93, 116
Stress, plane, 118, 119, 124, 125, 127, 132, 138, 142
Stress concentration, 4, 5, 8
Stress history, 54, 57, 63, 66
Stress point, 58, 59, 61, 67, 69, 71, 73, 76, 78

Stress profile, 96, 97, 99, 102, 107
Stress-strain curve, 52
Strip, 7, 21, 23
  eccentric hole in, 7
  orthotropic, 23
  semicircular notch in, 7
Symmetrization, 34
Symmetry, axial, *see* Axially symmetric
  problems
  orthorhombic, 26

Tension test, 52, 64, 65
Thermoelastic problem, 26, 27, 28, 29
Torsion, 119, 124, 127, 137
  of shafts of variable diameter, 24
  anisotropic, 24
Transformation, conformal, 5, 8
  linear, 25
Tresca yield condition, 96, 114, 118, 120, 127
Tube, 23
  thick-walled, 109, 120, 127, 135, 137, 141
  thin-walled, 120, 125, 135

Uniqueness, 53, 62, 87, 89, 91, 124
Unloading, 52, 64

Variational principles, *see* Minimum
  principles
Virtual work, principle of, 82

Wave, 36
  elastic, 28
  equation, 39, 40, 41
  forced, 36, 38
Wedge region, 20, 23, 24
Work, 59

Yield condition, 57, 58, 59, 61, 63, 66, 69, 71, 73, 75, 77, 113
  Mises, 127
  singular, 60, 72, 76, 85
  Tresca, 96, 114, 118, 120, 127
Yield frame, *see* Yield condition
Yield function, *see* Yield condition
Yield stress, 52, 56, 57, 64, 65
Yield surface, *see* Yield condition
Young's modulus, 56